Induction and Segregation of the Vertebrate Cranial Placodes

Developmental Biology

Editors

Daniel S. Kessler, *University of Pennsylvania School of Medicine*

Developmental biology is in a period of extraordinary discovery and research in this field will have a broad impact on the biomedical sciences in the coming decades. Developmental Biology is interdisciplinary and involves the application of techniques and concepts from genetics, molecular biology, biochemistry, cell biology, and embryology to attack and understand complex developmental mechanisms in plants and animals, from fertilization to aging. Many of the same genes that regulate developmental processes underlie human regulatory gene disorders such as cancer and serve as the genetic basis of common human birth defects. An understanding of fundamental mechanisms of development is providing a basis for the design of gene and cellular therapies for the treatment of many human diseases. Of particular interest is the identification and study of stem cell populations, both natural and induced, which is opening new avenues of research in development, disease, and regenerative medicine. This eBook series is dedicated to providing mechanistic and conceptual insight into the broad field of Developmental Biology. Each issue is intended to be of value to students, scientists and clinicians in the biomedical sciences.

Induction and Segregation of the Vertebrate Cranial Placodes
Byung-Yong Park and Jean-Pierre Saint-Jeannet
www.morganclaypool.com

ISBN: 9781615041022 paperback

ISBN: 9781615041039 ebook

DOI: 10.4199/C00014ED1V01Y201007DEB003

A Publication in the Morgan & Claypool Life Sciences Publishers series

DEVELOPMENTAL BIOLOGY

Book #3

Series Editor: Daniel S. Kessler, Ph.D., University of Pennsylvania

Series ISSN Pending

Induction and Segregation of the Vertebrate Cranial Placodes

Byung-Yong Park
Chonbuk National University, College of Veterinary Medicine
University of Pennsylvania, School of Veterinary Medicine

Jean-Pierre Saint-Jeannet
University of Pennsylvania, School of Veterinary Medicine

DEVELOPMENTAL BIOLOGY #3

MORGAN & CLAYPOOL LIFE SCIENCES

ABSTRACT

During evolution the vertebrate head has acquired a number of unique features including specialized paired sense organs and cranial sensory ganglia. These evolutionary novelties arise from discrete thickenings of the head ectoderm known as cranial placodes. They include the adenohypophyseal, olfactory, lens, trigeminal, profundal, otic, epibranchial and lateral line placodes. While distinct in the derivatives and cell types they will form, all cranial placodes originate from a common pre-placodal domain surrounding the anterior neural plate. Recent evidence suggests that the induction of this pre-placodal domain and its subsequent subdivision into individual placodes with specific identities is a multi-step process. Here we describe the development of these placodes and their derivatives and summarize recent advances in the characterization of the repertoire of transcription factors underlying their development. We also review recent studies that have started to address the role of several classes of signaling molecules in placode induction and segregation, including Bone Morphogenetic Proteins, Fibroblast Growth Factors and Wnt molecules.

KEYWORDS

cranial placode, otic, lens, olfactory, adenophypophyseal, lateral line, profundal, trigeminal, epibranchial, sensory, transcription factor, Six, Eya, Pax, Bmp, Fgf, Wnt

Contents

Introduction

During evolution the vertebrate head has acquired a number of unique characteristics, including an advanced craniofacial skeleton and specialized paired sensory organs. These important innovations accompanied the transition of vertebrate ancestors from small filter feeders to large active predators (Northcutt and Gans, 1983). Head skeleton and sensory ganglia originate from two embryonic structures: the neural crest and the cranial placodes. Neural crest and placodes share a number of important features pointing to a possible common evolutionary origin (Schlosser, 2008). They both arise from the neural plate border, boundary between the non-neural ectoderm and the neural plate; they delaminate from the epithelial structure from which they originate; and they have the ability to differentiate into a large array of cell types including sensory neurons, neuroendocrine cells, glia and supporting cells (Baker and Bronner-Fraser, 2001). However, the neural crest has also a set of very unique characteristics that clearly separates it from cranial placodes. Neural crest will develop a much broader repertoire of cell types compared to placodes, including pigment cells, cartilage and smooth muscle cells, and will migrate over greater distances in the embryo. Moreover unlike cranial placodes neural crest are not restricted to the head region, they arise from the entire length of the neural tube starting from a region posterior to the prospective diencephalon. Additionally, while neurons of cranial ganglia have a mixed origin from both neural crest and placodes, the peripheral ganglia in the trunk are exclusively neural crest derived (reviewed in Le Douarin et al., 1986; Baker and Bronner-Fraser, 2001).

Ectodermal placodes are transient thickenings of the embryonic head ectoderm. The term placodes also applies to developing organs such as teeth, mammary glands, hair follicles, feathers and scales, however here we will deal only with placodes that develop in the head and that form part of the sensory nervous system. This includes the adenohypophyseal, olfactory, lens, trigeminal, profundal, otic, epibranchial and lateral line placodes. In most vertebrates development of cranial placodes is initiated shortly after gastrulation when a pre-placodal field of naive ectoderm is established at the border between the future epidermis and the anterior neural plate/neural crest forming regions. According to their location along the anteroposterior axis and the influence of neighboring tissues each placode will adopt a specific identity (Figure 1). It is also important to

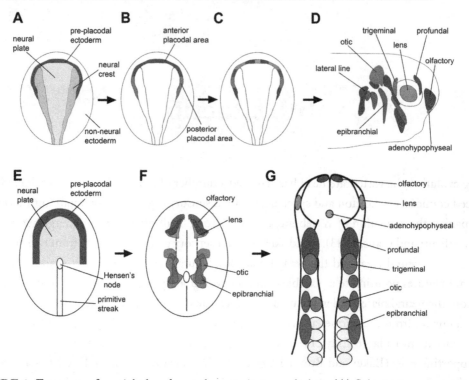

FIGURE 1: Fate map of cranial placodes at the anterior neural plate. (A) Schematic representation of a Xenopus laevis embryo at the neurula stage. The pre-placodal ectoderm abuts the neural plate anteriorly and is located lateral to the neural crest. (B) The pre-placodal ectoderm can be subdivided into two distinct domains the anterior and the posterior placodal areas which each give rise to a different subset of placodes. (C) A detailed fate map has not been established, however gene expression profiles predict that cranial placodes arise as illustrated in the diagram. Adenohypohyseal (light blue), olfactory (dark blue), lens (orange), profundal and trigeminal (green) placodes arise from the anterior placodal area. Otic and epibranchial (red and purple) and lateral line (brown) placodes are derived from the posterior placodal area. (A–C) Dorsal views, anterior to top. (D) Position of the cranial placodes in a tailbud stage Xenopus embryo Modified from Schlosser and Northcutt (2000). Lateral view, anterior to right, dorsal to top. Individual placodes and derivatives are color-coded as in panel (C). (E) Schematic representation of a chick embryo at the early neurula stage showing the position of the pre-placodal ectoderm, which delineate a horseshoe-shaped domain around the anterior neural plate. (F) Fate map of a 0–1 somite stage chick embryo shows that precursors for different placodes, olfactory/lens and otic/epibranchial, are intermingled and surround the neural plate. Based on data from Streit (2002) and Battacharyya et al., (2004). There is no fate map for the trigeminal placode, whose precursors presumably lie between the future olfactory/lens and otic/epibranchial domains. (G) At the 10–13 somite stage, placodes and their derivatives occupy distinct positions along the neuraxis. Modified from D'Amico-Martel and Noden (1983). (E–G) Dorsal views, anterior to top.

emphasize that placodes are not merely recipients of inducing signals, and once specified, they are in turn essential for normal development of surrounding structures. For example, the lens is essential for the normal development of the adjacent structures, the retina, iris and overlying cornea; the olfactory placode is required for normal forebrain development; and the otic epithelia induce chondrogenesis in the surrounding mesenchyme, which will provide the protective and structural capsule to the inner ear.

Cranial placodes include rather different structures, probably with different evolutionary origins. While some sensory placodes (otic and olfactory) may have homologues in basal chordates (Wada et al., 1998), the so-called neurogenenic placodes (trigeminal, otic, lateral line and epibranchial placodes) appear to have emerged at a later time (Shimeld and Holland, 2000). Generation of individual neurons from the ectoderm is a character shared by primitive chordates, therefore it is not the neurogenic potential of the ectoderm that is a vertebrate novelty, but more the concentration of foci of neurogenesis (ganglia) in discrete regions of ectoderm (Shimeld and Holland, 2000).

With the exceptions of the lens and the adenohypophyseal placodes, all cranial placodes will give rise to neurons in addition to other cell types. Neurogenic placodes contribute sensory neurons to cranial ganglia and may themselves be divided into two groups based on their location and fate (reviewed in Webb and Noden, 1993; Northcutt, 1996). Dorsolateral placodes (trigeminal, otic and lateral line) occupy relatively dorsal positions adjacent to the hindbrain and give rise to sensory cells of the inner ear and lateral line system, to the sensory neurons of the otic and lateral line ganglia supplying them, and to neurons of the trigeminal ganglion (among other cell types). Epibranchial placodes (geniculate, petrosal, and nodose) lie more ventrally, and are associated with the dorsocaudal aspect of the pharyngeal clefts and give rise to viscerosensory neurons of the facial, glossopharyngeal, and vagal nerves.

In recent years the development of the cranial placodes has been the object of a renewed interest, coinciding with the characterization of new molecular tools to identify placodes at various stages of development. This effort has led to a better understanding of their induction and diversification at the cellular and molecular levels, and has helped establish a draft of the gene regulatory network underlying placode formation. In the last 10 years a number of excellent reviews have discussed and reviewed some of the progress made in the field (Baker and Bronner-Fraser, 2001; Begbie and Graham, 2001; Streit, 2004; 2007; Battacharrya and Bronner-Fraser, 2004; Schlosser and Ahrens, 2004; Brugmann and Moody, 2005; Schlosser, 2005; 2006; 2007; 2008; Bailey and Streit, 2006; McCabe and Bronner-Fraser, 2009; Ladher et al., 2010). Here after a general overview of the development of the different placodes and their derivatives, we will summarize recent advances in characterization of the repertoire of transcription factors underlying placode development. We will also review recent studies beginning to address the role of several classes of signaling molecules in the induction and segregation of cranial placodes from a common pre-placodal region.

Cranial Placodes and Their Derivatives

The cranial placodes are localized ectodermal thickenings that develop by apicobasal elongation of cuboidal cells in the inner layer of the ectodem in the head of vertebrate embryos. They are involved in formation of sense organs (eye, nose, ear) and cranial sensory ganglia. Despite being grouped under the same term of "ectodermal placodes," and developing from the same initial territory, they differ greatly in their pattern of development as well as in the derivatives and cell types they generate. There are two major ways in which the placodal ectoderm will be converted into a specific derivative: either by invagination of the thickened epithelium into a vesicle that will separate for the surface ectoderm, or by delamination of cells from the thickened ectoderm and reaggregation in

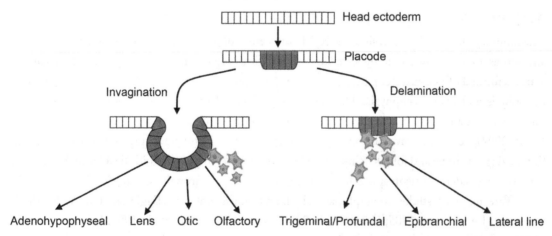

FIGURE 2: Schematic representation of the early morphogenetic processes associated with the development of the cranial placodes. All cranial placodes develop from a thickening of the head ectoderm. Adenohypophysis, olfactory epithelium of the nose, lens, and inner ear form by invagination of the placodal epithelium into a vesicle. In these placodes, with exception of the lens, cells will also delaminate from the invaginated structure to give rise to sensory neurons and secretory cells. Derivatives of the lateral line, epibranchial, trigeminal and profundal placodes form by delamination of cells from the placodal epithelium.

the underlying tissues (Figure 2). Delamination of cells is also occurring in placodal tissue forming by invagination with the exception of the lens. The resulting structures will undergo extensive proliferation and remodeling to give rise to very diverse cell types characteristic for each placodal derivative.

Six types of cranial placodes exist in higher vertebrates: 1) the adenohypophyseal placode forming the anterior lobe of the pituitary gland; 2) the olfactory placode that will give rise to the olfactory epithelium of the nose; 3) the lens placode that will differentiate into the transparent lens; 4) the trigeminal placode from which the sensory neurons of the ophthalmic and maxillomandibular lobes of the ganglion of the Vth cranial nerve originate; 5) the otic placode producing precursors for the sensory epithelia of the inner ear and neurons of the VIIIth cranial nerve; and finally, 6) the epibranchial placodes (geniculate, petrosal and nodose) that will produce the sensory neurons for the distal portion of the ganglia of the VIIth, IXth and Xth cranial nerves (the proximal neurons are derived from the neural crest). Fish and amphibians possess an additional placode, known as the lateral line, involved in the mechanosensory detection of water movements and electric fields. Finally, some amphibian species appear to have retained a primitive form of sensory placode known as the hypobranchial placode. Here we provide an overview of the development of each placode and the derivatives they will form.

ADENOHYPOPHYSEAL PLACODE

The pituitary gland or hypophysis is the regulatory center of growth, metabolism and reproduction, and acts as a relay between the hypothalamus and various organs. In most vertebrates the pituitary gland is formed of two distinct parts, an anterior and a posterior lobe known respectively as adenohypophysis and neurohypophysis. The adenohypophyseal placode forms the anterior lobe of the pituitary gland and gives rise to the endocrine secretory cells of the pituitary (reviewed in Asa and Ezzat, 2004), containing five hormone producing cell types: lactotropes (prolactin), gonadotropes (luteinizing hormone and follicle stimulating hormone), thyrotropes (thyroid stimulating hormone), corticotropes (adenocorticotropic hormone) and somatotropes (growth hormone).

Fate-mapping studies in amphibian, chick and mouse embryos (Eagleson et al., 1986; 1995; Couly and Le Douarin, 1985; Cobos et al., 2001; Osumi-Yamachita et al., 1994; Kawamura et al., 2002) have shown that the cells contributing to the adenohypophysis develop at the midline of the anterior neural ridge, which delineates the rostral boundary of the neural plate, a region devoid of neural crest. The anterior neural ridge also gives rise to the olfactory placodes and some forebrain tissues including the olfactory bulbs (reviewed in Papalopulu, 1995). Ablation of this region in chick embryos at the 2-4 somite stage confirmed these lineage analyses as it prevented formation of Rathke's pouch and any further pituitary development (elAmraoui and Dubois, 1993). Upon head folding, the oral ectoderm cells of the adenohypophyseal placode invaginate towards the prospective ventral diencephalon to form Rathke's pouch, the anlage of the adenohypophysis. Rathke's pouch starts as an invagination of the oral ectoderm in response to inductive signals from the prospective diencephalon. The region of the diencephalon above the pouch is known as the infundibulum and forms the posterior lobe of the pituitary or neurohypohysis (Figure 3). While in most basal fish and tetrapods the adenohypophyseal anlagen invaginates to form Rathke's pouch, in teleost fish the adenohypophyseal placode does not invaginate but rather maintains its initial organization forming a solid structure in the head (reviewed in Pogoda and Hammerschmidt; 2009).

The sequence of events leading to pituitary formation is especially well described at the cellular and molecular levels in the mouse embryo (reviewed in Burgess et al., 2002; Rizzoti and Lovell-Badge, 2005). In the mouse, the first sign of pituitary development occurs at 7.5 days post coitum (dpc) with the thickening of the ectoderm at the midline of the anterior neural ridge forming the adenohypophyseal placode. As development proceeds, the anterior neural tube bends and rapidly expands, displacing the adenohypophyseal placode ventrally, within the ectoderm at the roof of the

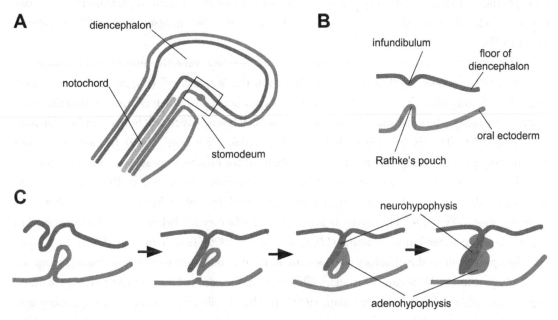

FIGURE 3: Development of the adenohypophysis. (A) Schematic representation of a sagittal section through the stomodeum of an early mouse embryo. The boxed area corresponds to the region from which the future adenohypophysis will develop as shown in subsequent panels. (B) Invagination of the oral ectoderm and evagination of the roof of the diencephalon give rise to Rathke's pouch and infundibulum, respectively. (C) Sequence of morphogenetic processes associated with the development of the adenohypophysis (Rathke's pouch) and neurohypophysis into the pituitary gland.

future oral cavity. Around 9 dpc, the placode forms an invagination, the rudimentary Rathke's pouch. Then, a restricted region of the prospective ventral diencephalon above the pouch gives rise to the infundibulum from which the posterior pituitary or neurohypohysis forms. The close juxtaposition of Rathke's pouch and the diencephalon is required for tissue interactions between neural and oral ectoderm. By 10.5 dpc, the pouch is fully developed, and at 12.5 dpc it is completely detached from the oral ectoderm and becomes intimately associated with the neurohypohysis. Further shaping and branching of Rathke's pouch into the mature adenohypohysis varies greatly between species.

OLFACTORY PLACODES

The olfactory system transduces signals from the outside world through a group of sensory neurons (olfactory receptor neurons) whose axons project directly into the olfactory bulb. The olfactory epithelium harboring these neurons line the dorsal roof, septum and lateral turbinates of the caudal region of the nasal cavity, and is derived from the olfactory placodes. As for the adenohypohyseal placode, classical transplantation and cell labeling experiments in the chick, *Xenopus* and zebrafish embryos indicate that the olfactory placodes arise from the anterior neural ridge, from two areas lateral to the adenohypohyseal anlagen (Couly and Le Douarin, 1985; Eaglson et al., 1995; Kozlowski et al., 1997).

In Xenopus as the neural plate closes to form the neural tube the anterior neural ridge coalesces to form an outer epithelial structure known as the "sense plate" (Figure 4). At the tailbud stage the olfactory placodes start to thicken and differentiate as two bilateral areas within the sense plate. At this stage, the sense plate is composed of two cell layers, which will both contribute distinct cell types to the olfactory epithelium (Klein and Graziadei, 1983; Burd, 1999). In most vertebrates the olfactory placodes invaginate to form the epithelia of the olfactory organ, which is odor-sensing and the vomeronasal organ mainly used to detect pheromones (reviewed in Buck, 2000). Some amphibians develop an additional cavity, known as the middle cavity, which often occurs *de novo* at metamorphosis (Figure 4). This cavity is lined with an olfactory epithelium specifically involved in the detection of water-borne odorants (Higgs and Burd, 1999; Taniguchi et al. 2008). In all species the epithelium of the olfactory organs contain ciliated olfactory neurons, harboring receptors responding to odor-producing substances dissolved in the serous layer covering the epithelium, supporting cells and basal cells (Farbman, 1994). The basal cells are stem cells of the olfactory epithelium that have the ability to divide and continuously produce new neurons during embryogenesis and throughout the adult life of the organism. The axons of the sensory neurons project to the olfactory bulb forming the olfactory, vomeronasal and terminal nerves.

The olfactory placodes also give rise to glial cells that ingress and migrate along the olfactory nerve towards the brain (Ramon-Cueto and Avila, 1998). Olfactory placodes are the only ectodermal placodes to produce glia, a cell type typically derived from neural crest (Couly and LeDouarin,

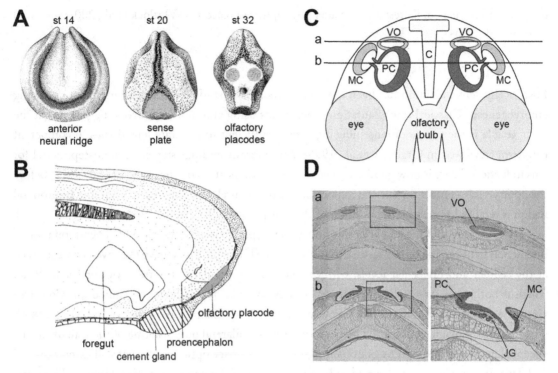

FIGURE 4: Development of the olfactory placode in the frog *Xenopus laevis*. (A) Frontal views of Xenopus embryos showing the position of the anterior neural ridge (stage 14), the sense plate (stage 20) and the olfactory placodes (stage 32). Modified from Drysdale and Elinson (1991). (B) Diagram of a longitudinal section of a stage 35 embryo at the level of the prosencephalon showing the position of the olfactory placode. Modified from Hausen and Riebesell (1991). (C) The diagram illustrates a longitudinal section of a tadpole head. The relative positions of the nasal cavities are indicated. Principal cavity (PC), middle cavity (MC) and vomeronasal organ (VO). Also shown are the cartilage separating the nasal capsules (C), the olfactory bulb with associated olfactory nerves and the eyes. Modified from Higgs and Burd (1999). The lines labeled "a" and "b" correspond to the level of the sections shown in the next panel. (D) Transverse sections at two different levels through the nasal cavity of a stage 55 Xenopus tadpole. The panels on the right are higher magnification views of the boxed area in the left panels. JG, Jacobson's gland.

1985; Baker and Bronner-Fraser, 2001). The olfactory placodes are also producing gonadotropin-releasing hormone (GnRH) neurons that migrate along the olfactory and vomeronasal nerves into the brain (reviewed in Wray, 2002). GnRH neurons project their axons in different regions of the brain, which have been implicated in the control of some aspects of reproduction. Recent lineage tracing experiments in zebrafish challenged this view, suggesting that GnRH neurons may in fact originate from the neural crest and the adenohypophyseal placode (Whitlock et al., 2003; reviewed in Whitlock, 2005).

LENS PLACODES

Historically, lens induction has been a preferred model for studying inductive interactions during embryogenesis. Classical transplantation experiments using amphibian embryos suggested that the optic vesicle is the source of lens-inducing signals sufficient to generate lens tissues in competent ectoderm (reviewed in Grainger et al., 1996). More recent findings suggest a multistep model for lens induction. There is now good evidence that lens specification occurs at the neurula stage, before the optic vesicle contact the surface ectoderm, and that neural crest cell migration in the frontonasal region is required to restrict the position of the lens placode (Bailey et al., 2006).

Development of the lens is closely linked to morphogenesis of the eye and depends on mutual interaction between the optic cup and the lens vesicle. These processes have been well characterized at the cellular and molecular levels, with one key master regulator, the transcription factor Pax6, which is both necessary and sufficient for eye development throughout the animal kingdom (reviewed in Ogino and Yasuda, 2000; Chow and Lang, 2001; Lang, 2004; Donner and Mass, 2004). The first morphological indication of eye formation is a bilateral outpocketing of the wall of the diencephalon. These outpocketings deepen to form the primary optic vesicle that will extend toward and contact the surface ectoderm to induce thickening of the prospective lens placode. The optic vesicle progressively bends around the lens placode, forming the bilayered optic cup. The optic cup will differentiate into two layers. The outer layer produces melanin pigment and ultimately becomes the retinal pigmented epithelium. Cells of the inner layer proliferate rapidly and generate a variety of ganglion cells, glia, interneurons and light sensitive photoreceptors neurons. Collectively, these cells constitute the neural retina. As the cavity of the optic cup deepens, the lens placode invaginates into the cup to form an open lens vesicle. Eventually the lens vesicle closes and breaks away from the superficial ectoderm to constitute a rounded epithelial body lying inside the optic cup (Figure 5).

Differentiation of the lens into a transparent structure with the appropriate optical qualities involves a complex sequence of events culminating in the synthesis of a class of lens-specific proteins known as crystallins. This process involves a regulatory pathway that is both complex and evolutionary conserved (reviewed in Reza and Yasuda, 2004; Lovicu and McAvoy, 2005). Initially the posterior cells of the lens elongate to form the primary lens fibers while the anterior cells retain a low cuboidal configuration. This process reduces the original cavity of the lens vesicle to a slit. The transformation from epithelial cells to lens fibers takes place within a domain at the equator of the lens. Adjacent to this domain is a germinative region of dividing cells. These cells progressively move into the equatorial zone where they stop dividing to initiate elongation and differentiation.

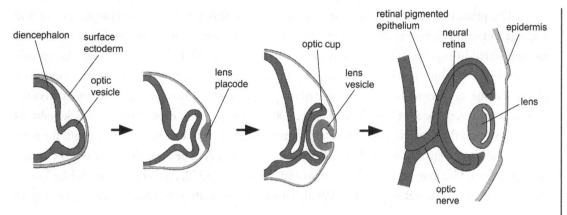

FIGURE 5: Development of the lens placode into the mature lens. Diagrams illustrating the sequence of events leading to formation of the lens in a vertebrate embryo. The presumptive lens placode ectoderm is in direct contact with the optic vesicle. The ectoderm thickens to form the lens placode. The lens placode then invaginates to form the lens vesicle. The optic vesicle bends around it to form the bilayered optic cup. The lens vesicle then buds off the surface ectoderm so that the mature lens ends up within the cup-shaped retina which is now composed of two layers, the retinal pigmented epithelium and the neural retina that generates the light sensitive photoreceptor neurons.

PROFUNDAL/TRIGEMINAL PLACODES

The trigeminal ganglion complex of cranial nerve V innervates much of the head. This ganglion develops from two separate ganglia, the ophthalmic and maxillomandibular (reviewed by Baker and Bronner-Fraser, 2001). In anamniotes, the ophthalmic and maxillomandibular lobes of the trigeminal complex are referred as the profundal and trigeminal placodes, respectively. In most organisms the two ganglia fuse during embryogenesis into a single unit. In *Xenopus*, the profundal and the trigeminal ganglia are separate distally but fused at their proximal end as they condense around stage 24 (Schlosser and Northcutt, 2000). In fish, frogs, birds, and mice the profundal/trigeminal or ophthalmic/maxillomandibular placodes contribute cutaneous sensory neurons to their respective ganglia in a similar manner (Schilling and Kimmel, 1994; Schlosser and Northcutt, 2000; Hamburger 1961; D'Amico-Martel and Noden, 1983; Ma et al., 1998). These neurons extend peripheral axons underneath the skin of the head, to detect mechanical, chemical, and thermal stimuli, and central axons into the hindbrain, to communicate these inputs to the central nervous system. In mammals, the ophthalmic branch of the trigeminal ganglion complex innervates the skin of the head region, the eyeball and eye muscles, and the nose; the maxillary branch innervates the upper jaw, while the mandibular branch innervates the lower jaw and tongue (Baker and Bronner-Fraser, 2001).

The profundal/trigeminal placodes are positioned halfway between the prospective eye and ear, adjacent to the future midbrain-hindbrain boundary. Factors secreted by the dorsal neural tube are implicated in trigeminal placode induction (Stark et al., 1997; Baker et al., 1999). As recently described in zebrafish the sensory neurons are typically born as scattered cells delaminating from placodal tissue and accumulating in the underlying mesenchyme. These neurons eventually coalesce to form a compact cluster of cells (Knaut et al., 2005). Transplantion experiments in chick embryos have shown that the trigeminal ganglion has a mixed origin. It contains neurons from both the neural crest and from placodes, with glial cells and all supporting cells entirely derived from the neural crest (Figure 6; Hamburger, 1961; Narayanan and Narayanan, 1978; Ayer-Le Lievre and Le Douarin 1982; D'Amico-Martel and Noden, 1983). This dual contribution extends to most organisms. In

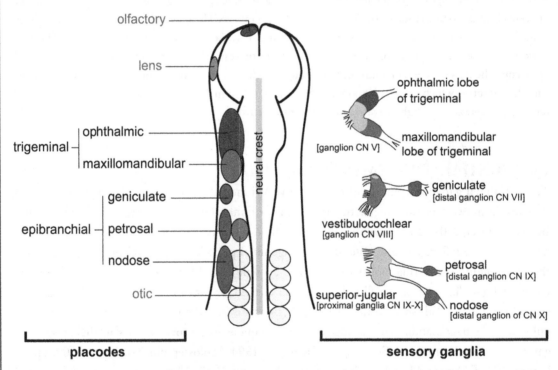

FIGURE 6: Trigeminal and epibranchial placodes contribution to sensory neurons in the chick embryo. Sensory ganglia and the placodes from which they arise are color-coded on this diagram of a chick embryo. The position of the trigeminal and epibranchial placodes is shown on the left side. Corresponding sensory ganglia to which they contribute are depicted on the right. The neural crest (yellow) contributes neurons to the proximal aspect of the trigeminal ganglion complex of cranial nerve V (CN V), and the distal aspect of cranial nerves VII (CN VII), IX (CN IX) and X (CN X). Modified from D'Amico-Martel and Noden (1983) and Baker and Bronner-Fraser (2001).

Xenopus, neural crest cells appear to join the cells aggregating under the placodal region at stage 21, however there is no direct evidence that this neural crest contribution applies equally to both ganglia of the trigeminal complex (Schlosser and Northcutt, 2000).

OTIC PLACODES

The inner ear is the organ responsible for hearing, balance and detection of acceleration, and it is almost entirely derived from the otic placode. With the exception of the pigment cells of the stria vascularis and the secretory epithelium of the cochlea, which are of neural crest origin, all components of the inner ear derive from the otic placode (reviewed in Torres and Giraldez, 1998; Fekete and Wu, 2002; Noramly and Grainger, 2002; Whitfield et al., 2002; Riley and Phillips, 2003, Barald and Kelley, 2004).

In most species the thickening of the ectoderm into a placode occurs in a region adjacent to rhombomere 5 (reviewed in Ohyama et al., 2007), while in amphibians the otic placode is centered onto rhombomere 4 (Ruiz i Altaba and Jessell, 1991). Transplantation and ablation experiments suggest that otic placode induction depends on signals derived from surrounding tissues, the prospective hindbrain and the head mesoderm. However, the relative importance of these inducing tissues differs from one species to another (reviewed in Jacobson, 1966; Torres and Giráldez, 1998; Baker and Bronner-Fraser, 2001; Noramly and Grainger, 2002; Riley and Philipps, 2003; Ohyama et al., 2007). For example, in zebrafish grafting hindbrain tissue on the ventral side of an embryo induces ectopic otic vesicles (Woo and Fraser, 1998). Mutations disrupting formation of the prechordal plate and paraxial head mesoderm delay, but do not prevent otic placode induction in zebrafish (Mendonsa and Riley, 1999). In the chick, removal of paraxial head mesoderm underlying the presumptive otic ectoderm prevents otic placode development, even when replaced by mesoderm from a different origin (Kil et al., 2005). In *Xenopus* prospective hindbrain-derived signals are sufficient to initiate otic development in the absence of mesoderm cues (Park and Saint-Jeannet, 2008).

Once specified, the otic placode invaginates to form an otic cup, which eventually separates from the surface ectoderm to form the otic vesicle or otocyst, a rounded structure without apparent polarity (Figure 7). As the otic placode invaginates into a cup neuroblasts delaminate from the anterior ventral aspect of the otic epithelium to give rise to neurons of the vestibulocochlear (statoacoustic) ganglion of cranial nerve VIII. This ganglion will provide afferent innervation for hair cells associated with both the auditory and vestibular components of the inner ear (Figure 7). Differential rates of cell division and complex morphogenetic movements will shape the otic vesicle into a highly specialized and asymmetrically organized structure.

The vertebrate otocyst can be divided into two functional domains along its dorsoventral axis: the ventral region is responsible for the sense of hearing, and the dorsa region is involved in vestibular functions. In the mature inner ear, the utricle and semicircular canals constitute the

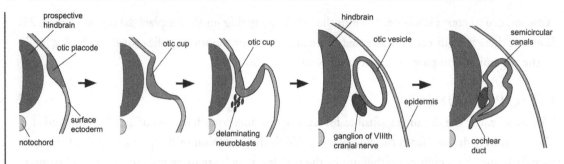

FIGURE 7: Development of the otic placode into the inner ear. Diagrams illustrating the development of the inner ear in a vertebrate embryo. Formation of the otic placode by thickening of the ectoderm adjacent to the prospective hindbrain. The placode invaginates to form first the otic cup and then the otic vesicle or otocyst. During that process, individual neuroblasts delaminate from its anteroventral region and coalesce to form the vestibulochochlear ganglia of cranial nerve VIIIth. The otocyst will go through differential growth and extensive remodeling to give rise to the fluid-filled labyrinth of the mature inner ear. The fully developed mammalian inner ear can be divided into two functional domains: ventrally the cochlear duct responsible for the sense of hearing, and dorsally the semicircular canals involved in the vestibular functions of the inner ear.

vestibular apparatus. Structurally and functionally these elements are highly conserved in vertebrates. In contrast, the ventrally located auditory chambers have undergone more extensive evolutionary modifications. The saccule and lagena are prominent auditory organs in fish but the saccule has a vestibular role in mammals and birds, and the lagena is absent in mammals. The primary auditory organ in mammals and birds is the cochlea, which has no known counterpart in amphibians and fish (Riley and Phillips, 2003). In amphibians, the saccule, the basilar papilla and the amphibian papilla assume auditory function. In this organism the saccule is a low-frequency vibration/sound detector, while the amphibian papilla and basilar papilla, are low- to mid-frequency and high-frequency sound detectors, respectively (Smotherman and Narins, 2000).

Each chamber is associated with a sensory epithelium containing support cells and mechanosensory hair cells, which convey auditory and vestibular information. The maculae are the sensory epithelia of the utricle (utricular macula) and saccule (saccular macula). The sensory epithelia associated with each semicircular canal, known as anterior, horizontal and posterior cristae, are sensitive to fluid motion caused by angular acceleration. The sensory epithelia in the utricle, saccule and lagena, are associated with otoliths or otoconia (small crystals of calcium carbonate), which facilitate vestibular and auditory function by transmitting accelerational forces and sound vibrations, respectively, to mechanosensory hair cells (reviewed in Fritzsch et al., 2002; Barald and Kelley, 2004; Fritzsch et al., 2006).

EPIBRANCHIAL PLACODES

The three epibranchial placodes are located ventral to the otic placode, and dorsocaudal to the pharyngeal clefts, which separate each pharyngeal (branchial) arch. Epibranchial placodes-derived neurons innervate internal organs to transmit information such as heart rate, blood pressure, and visceral distension from the periphery to the central nervous system (Baker and Bronner-Fraser, 2001). From rostral to caudal the epibranchial placodes comprise the geniculate, petrosal, and nodose placodes, each associated in sequence with the first, second and third branchial clefts. Each placode contributes sensory neurons to cranial nerves VII (facial nerve), IX (glossopharyngeal nerve), and X (vagal nerve), respectively (Figure 6). More specifically, epibranchial placodes contribute viscerosensory neurons solely to the distal ganglia of cranial nerves VII, IX and X, innervating several visceral organs and the taste buds (reviewed in Northcutt, 2004). The proximal ganglia of cranial nerves VII, IX and X are derived from neural crest and produce somatosensory neurons (Narayanan and Narayanan, 1980; Ayer-Le Lievre and Le Douarin; 1982; D'Amico-Martel and Noden, 1983). While the distal (placodally-derived) and proximal (neural crest-derived) ganglia of cranial nerves VII, IX, and X are clearly distinct in amniotes, they cannot be distinguished in *Xenopus* due to the early fusion of the ganglion primordia (Schlosser and Northcutt, 2000). In a very detailed analysis, Schlosser and Northcutt have shown that during *Xenopus* development, the facial ganglion is intimately fused with the anteroventral lateral line ganglion, while the glossopharyngeal ganglion is fused with the middle lateral line ganglion (Schlosser and Northcutt, 2000). Fish and amphibians form additional vagal epibranchial placodes associated with more posterior branchial clefts. For example three vagal epibranchial placodes have been described in Xenopus (Schlosser and Northcutt, 2000) and six in the lamprey (Damas, 1951).

Differentiation occurs by delamination of cells from the thickened placode, similar to what is seen for the trigeminal and lateral line placodes (Figure 2). Delaminated neuroblasts then coalesce to form ganglia that make appropriate connections to their targets. Early work in the chick embryo suggested that these placodes were induced by signals derived from adjacent tissues including the cranial neural crest and the pharyngeal endoderm (Webb and Noden, 1993). However, later ablation experiments have shown that epibranchial placodes can form in the absence of neural crest (Begbie et al., 1999). Recent work in zebrafish indicates that epibranchial placodes induction depends on cranial mesoderm-derived signals that establish both the epibranchial placodes and development of the pharyngeal endoderm, which is subsequently required to promote neurogenesis in the epibranchial placodes (Nechiporuk et al., 2007). Another study suggested that the initial induction of both otic and epibranchial placodes share common signals derived from the hindbrain (Sun et al., 2007).

Development of the otic and epibranchial placodes was long thought to be independent, consistent with the very distinct functions mediated by the inner ear and the epibranchial ganglia.

Recent studies indicate that may not be the case. For example cells destined to form the otic and epibranchial placodes are initially intermingled within the pre-placodal ectoderm (Streit, 2002). In addition, recent molecular analyses suggest that epibranchial and otic placodes originate from a common precursor domain defined by Pax2 expression (Sun et al., 2007; reviewed in Ladher et al., 2010). In fact, these properties are also shared by the lateral line placode, suggesting that all three placodes might be developmentally and evolutionary related (Baker et al., 2008).

LATERAL LINE PLACODES

The lateral line system is an important mechanosensory organ in fish and amphibians and is involved in the detection of water motion and electric fields. It allows for predator and prey detection, object avoidance and social behaviors, such as schooling and sexual courtship (Dijkgraaf, 1963; Montgomery et al., 2000). The lateral line system has completely disappeared in terrestrial tetrapods. The name of "lateral line" comes from the fact that the sensory organs are arranged in lines along each flank on the surface of the body. The lateral line receptor organs include mechanoreceptive neuromasts that respond to disturbances in the water, and in some cases electroreceptive organs that respond to weak electric fields (Webb, 1989). Neuromasts contain a core of mechanosensory hair cells, surrounded by support cells, and are innervated by sensory neurons located in a cranial ganglion (Figure 8A). The neuromasts on the head form the so-called anterior lateral line system, the ganglion of which is located between the ear and the eye, while the neuromasts on the body and tail, including those on the caudal fin, form the posterior lateral line system, its ganglion being just posterior to the ear (Figure 8B). Neuromasts are located either superficially, in shallow dermal pit

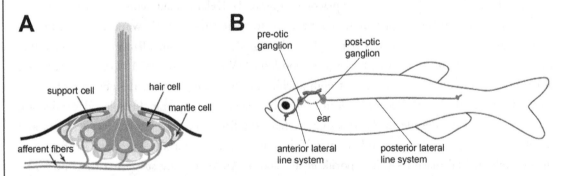

FIGURE 8: Organization of the zebrafish lateral line system. (A) Schematic representation of a neuromast. The mechanosensory hair cells are surrounded by support and mantle cells. (B) Schematic representation of an adult fish, illustrating the innervation of the anterior (red) and posterior (green) lateral line systems. The sensory neurons for each lateral line system are gathered in a pre-otic and a post-otic ganglion, respectively. Modified from Ghysen and Dambly-Chaudiere (2004).

lines, in cartilaginous grooves, or embedded into bony or cartilaginous canals. Electroreceptors are found either in ampullary or tuberous structures. These ampullary organs are largely restricted to the head while tuberous organs are found in both the head and the trunk. Neuromasts receive afferent and efferent innervation, while electroreceptors are only supplied by afferent fibers (reviewed in Ghysen and Dambly-Chaudiere, 2004; Gibbs, 2004).

Comparative studies of lateral line systems suggest that the lateral line arises from six pairs of lateral line placodes in fish, three located anterior to the otic vesicle (anterodorsal, anteroventral and otic lateral line placodes) and three posterior to the otic vesicle (middle, supratemporal and posterior lateral line placodes). Most amphibians have only five lateral line placodes, due to the loss of the otic lateral line placode, and completely lack electroreceptors (Schlosser and Northcutt, 2000; reviewed in Schlosser, 2002a; 2002b). Each lateral line placode gives rise to a single lateral line nerve that will innervate lateral line hair cells and convey information to the adjacent hindbrain, and to one migrating primordium, which extends and migrates to form the lateral line systems of the head and trunk.

The initial stages of lateral line development from the corresponding placodes have been relatively well described in fish and amphibians (Winklbauer, 1989; Northcutt et al., 1994; Northcutt, 1997; Ledent, 2002; Sarrazin et al., 2010). Briefly, cells delaminating from the thickened placodal epithelium coalese in the underlying mesenchyme to give rise to sensory neurons of the lateral line ganglia, while the remaining placodal cells become lateral line primordia. This primordium elongates or migrates between surface ectoderm and basement membrane in a stereotypical pattern depositing mechanosensory neuromasts at regular intervals. The growth cones of axons of the lateral line ganglion extend along with the primordium, innervating neuromasts as they are deposited. Cephalic lateral line placodes elongate at the rostral end of each placode, whereas the trunk lateral line placodes and their derivatives amigrate caudally along the trunk. The pattern of lateral line distribution and canal branching, and the number of lateral line receptors is species-specific (Northcutt, 1989).

In recent years the development of the lateral line system has been more extensively studied in the zebrafish, an organism highly suitable for live imaging (reviewed in Dambly-Chaudiere et al., 2004; Ghysen and Dambly-Chaudiere, 2004; 2007). More specifically, some of the most recent studies have focused on the signaling pathways (chemokine, fibroblast growth factor, and Wnt) regulating cellular behaviors associated with posterior lateral line migration (Dambly-Chaudiere et al., 2007; Aman and Piotrowski, 2008; Nechiporuk and Raible, 2008; reviewed in Ma and Raible, 2009; Aman and Piotrowski, 2009).

HYPOBRANCHIAL PLACODES

A unique type of neurogenic placode known as hypobranchial placodes was recently reported in some amphibian species (Schlosser et al., 1999; Schlosser and Northcutt, 2000; Schlosser, 2003). In *Xenopus*, these placodes are located ventral and caudal to the second and third pharyngeal pouches and give rise to small hypobranchial ganglia of yet unknown function. They have also been found in another anuran, the direct developing frog *Eleutherodactylus coqui* (Schlosser, et al., 1999; Scholsser, 2003). Nothing is known about the mechanisms regulating the development of these placodes. Because the development of both epibranchial and hypobranchial placodes arise from a larger placodal domain in the branchial region, it is hypothesized that these newly identified placodes may correspond to ventral extensions of epibranchial placodes (Schlosser, 2003). Neurogenic hypobranchial placodes have not been reported in chick, mouse or zebrafish embryos, raising the possibility that they represent primitive vertebrate characteristics that may have been lost during evolution.

Molecular Identity of Cranial Placodes

All vertebrate cranial placodes originate from a common territory known as the pre-placodal or pan-placodal ectoderm, adjacent to the anterior neural plate (Figure 1). Studies in the chick embryo using focal dye labeling have shown that within the pre-placodal region, precursors for different placodes are initially intermingled, still some separation of individual populations along the anterior posterior axis is already apparent at the end of gastrulation (Streit, 2002; Bhattacharyya et al., 2004). Precursors for anterior placodes (adenohypophysis, olfactory, lens) are located in the rostral most pre-placodal region, while precursors for posterior placodes (trigeminal, epibranchial, otic, lateral line) are restricted more caudally (Figure 1; D'Amico-Martel and Noden, 1983; Kozlowski, et al., 1997; Streit, 2002; Bhattacharyya, et al., 2004). This subdivision can be visualized by the regional expression of transcription factors, shortly after the induction of the pre-placodal domain. As development proceeds, the pre-placodal region becomes molecularly divided into smaller sub-domains such that by the time placodes can be identified morphologically, each appears to express a unique combination of transcription factors - a transcriptional code that may underlie their identity. This is illustrated in Figure 9, for the expression of a several transcription factors during cranial placode development in *Xenopus*. These genes include the transcription factors Six1 (Pandur and Moody, 2000), initially broadly expressed in the entire pre-placodal region; Pax8 (Heller and Brandli, 1999), restricted to a posterior sub-domain of the pre-placodal region, corresponding to precursors of the otic and lateral line placodes; Dmrt4 (Huang et al., 2005), confined anteriorly to the prospective regions of the adenohypophyseal and olfactory lens placodes; Foxi1c (Pohl et al., 2002) expressed on the anlagen of the epibranchial and lateral line placodes; Pax6 (Hirsch and Harris, 1997), expressed in the neural plate and in precursors of the olfactory and lens placodes; and Ngnr1 (Perron et al., 1999), expressed in the trigeminal/profundal placode (Figure 9A). Later in development and as the placodes segregate from one another, the expression of these genes may persist in various combinations in the placode derivatives, concomitant to the differential activation of novel genes, such as Tbx2 (Takabatake *et al.,* 2000) expressed in the otic and trigeminal/profundal placodes, and Lens1 (Kenyon *et al.,* 1999) restricted to the lens placode (Figure 9B–C).

The gene regulatory network underlying cranial placode development has been the focus of intense investigation in the last decade, and it is just beginning to be unraveled (reviewed in Baker and

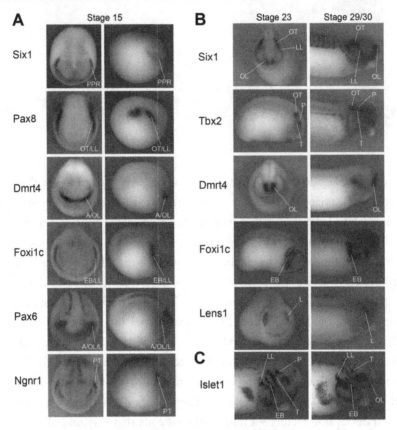

FIGURE 9: Whole-mount in situ hybridization of several transcription factors differentially expressed in the Xenopus laevis cranial placodes. (A) Expression of genes in the pre-placodal domain at the early neurula stage (stage 15). Left panels are anterior views, dorsal to top. Right panels are lateral views, dorsal to top, anterior to right. Six1 is initially broadly expressed in the entire pre-placodal region (PPR). Pax8 is restricted to a posterior sub-domain of the pre-placodal region, corresponding to precursors of the otic (OT) and lateral line (LL) placodes. Pax8 is also detected in the developing pronephros. Dmrt4 is confined anteriorly to the prospective regions of the adenohypophyseal (A), and olfactory (OL) placodes. Foxi1c is restricted to the epibranchial (EB) and lateral line (LL) placodes. Pax6 is expressed in the posterior neural plate but also in precursors of the adenohypophyseal (A), olfactory (OL) and lens (L) placodes. Ngnr1 has multiple domains of expression among which it is detected in the profundal/trigeminal (PT) placode. (B) Later in development and as the various placodes segregate and aquire distinct identities they will express different combination of transcription factors. Left panels (stage 23), are frontal views, dorsal to top, except for Tbx2 and Foxi1c, which are lateral views, dorsal to top, anterior to right. Left panels (stage 29/30) are lateral views, dorsal to top, anterior to right. (C) Islet1 expression in the cranial region of a stage 29/30 (left panel) and a stage 35/36 (right panel) embryos. Lateral views, dorsal to top, anterior to right. The embryonic stages are according to Niewkoop and Faber (1967).

Bronner-Fraser, 2001; Streit, 2004; Battacharrya and Bronner-Fraser, 2004; Schlosser and Ahrens, 2004; Brugmann and Moody, 2005; Schlosser, 2006; Bailey and Streit, 2006; McCabe and Bronner-Fraser, 2009). In the following tables we summarize the expression of transcription factors during the development of the seven cranial placodes and their derivatives. These tables will compare data from multiple vertebrate species, including fish, frog, chick and mouse, as follow: adenohypophyseal placode (Table 1), olfactory placode (Table 2), lens placode (Table 3), trigeminal/profundal placode (Table 4), otic placode (Table 5), epibranchial placode (Table 6) and lateral line placode (Table 7). For each gene, the listed references report the expression pattern, and when available, the function of the gene in the corresponding placode. These tables are not intended to be an exhaustive compilation of all the transcription factors within the cranial placodes and their derivatives but more a summary of the current knowledge as a way to gain insight into the molecular identity of these placodes.

A few years ago, Meulemans and Bronner-Fraser (2004), presented a comprehensive analysis of the regulatory network of genes expressed at the neural plate border, which includes neural crest and placodes. In this model they proposed that in response to signaling events a number of transcriptions factors are sequentially induced at the neural plate border in a two-step process. First, a group of genes is activated, referred as "neural plate border specifiers," which includes members of the Zic, Pax, Dlx and Msx families of transcriptional regulators. These factors, which are broadly expressed at the neural plate border, are in turn responsible for the activation of a subset of genes with more restricted expression domains, known as "neural crest specifiers" (Meulemans and Bronner-Fraser, 2004) or "pre-placodal specifiers" (Litsiou et al., 2005). The function of these neural crest and pre-placodal specifiers is to control the expression of genes regulating the behavior and differentiation patterns of these two cell lineages. Significant advances have been made over recent years into the regulation of placode formation, which has helped in drafting a constantly evolving gene regulatory network underlying vertebrate placode development. Among the key regulators of placode development are the Six, Eya and Pax gene families, which are expressed by all vertebrate sensory placodes and constitute an important branch of this regulatory network. This specific topic has been reviewed in a number of excellent recent articles (Streit, 2004; 2007; Battacharrya and Bronner-Fraser, 2004; Schlosser and Ahrens, 2004; Brugmann and Moody, 2005; Schlosser, 2006; 2007; Bailey and Streit, 2006; McCabe and Bronner-Fraser, 2009), and will be only briefly summarized here.

Six AND Eya GENE FAMILIES

In vertebrates, the *Six* and *Eya* gene families have six (Six1–6) and four (Eya1–4) members, respectively. *Six* genes encode transcription factors with direct DNA-binding capacity, while *Eya* genes encode nuclear proteins with tyrosine phosphatase activity. Eya proteins affect transcription indirectly by interaction with other proteins including Six transcription factors (reviewed in Kawakami et al.,

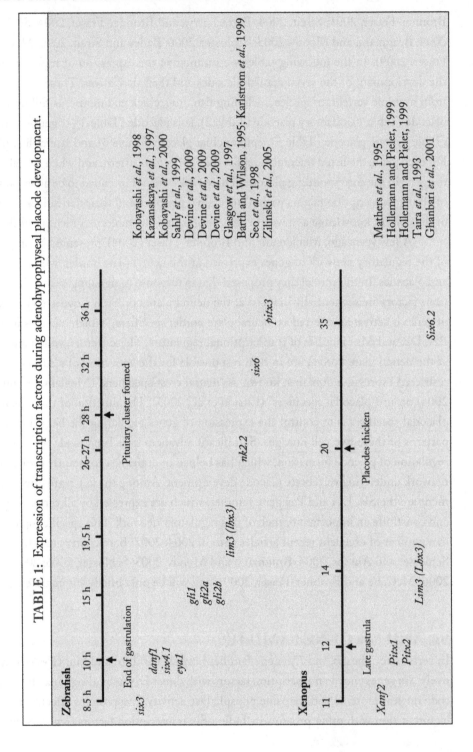

TABLE 1: Expression of transcription factors during adenohypophyseal placode development.

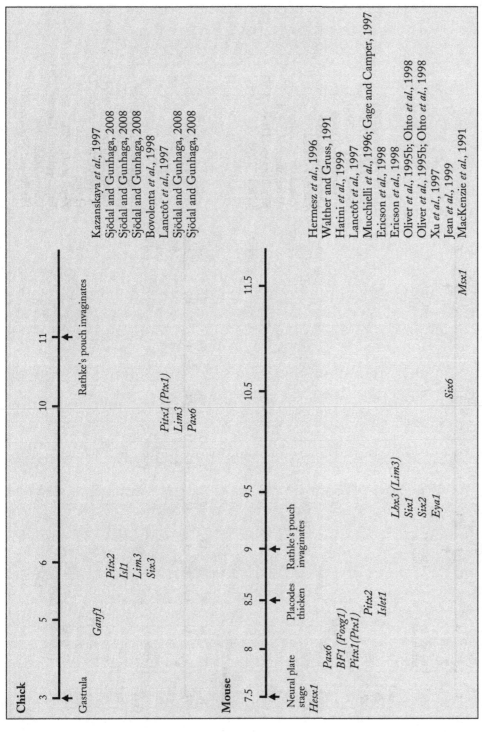

Note: For each table the time-line of development of zebrafish (hours), Xenopus (stages) and chick (stages) embryos are according to Kimmel et al. (1995), Nieuwkoop and Faber (1963), and Hamburger and Hamilton (1992), respectively. For the mouse the stages are embryonic days post coitum. Alternate names for genes are indicated in parentheses.

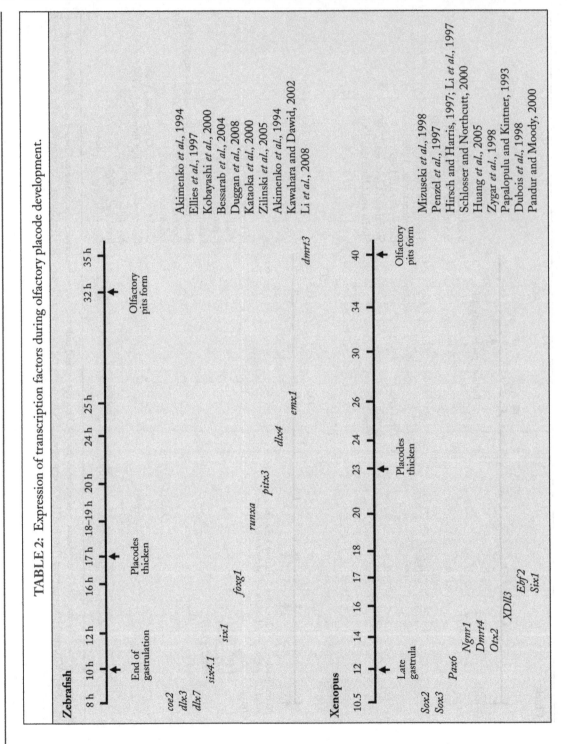

TABLE 2: Expression of transcription factors during olfactory placode development.

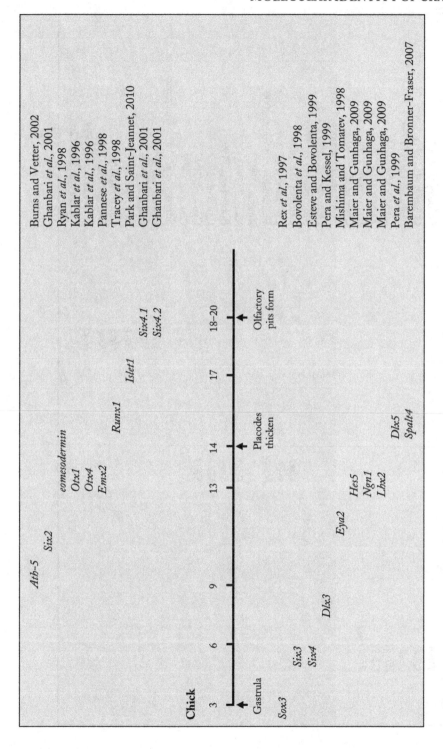

TABLE 2: (continued)

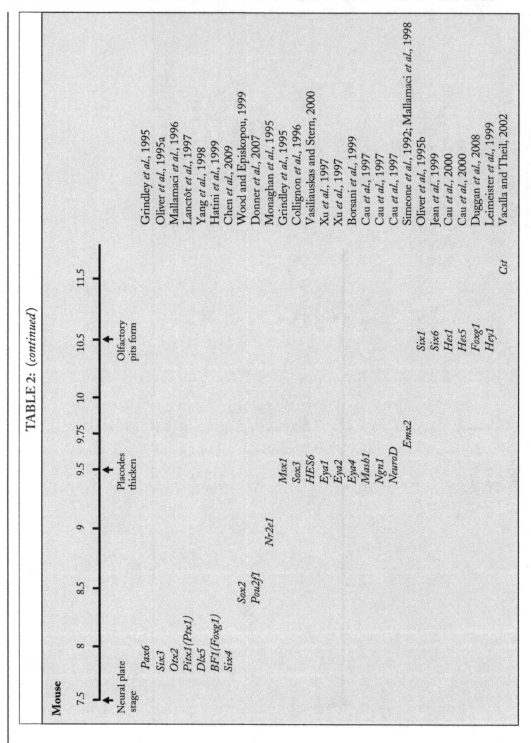

Mouse

	7.5	8	8.5	9	9.5	9.75	10	10.5	11.5	
Neural plate stage	↑				Placodes thicken ↑			Olfactory pits form ↑		
		Pax6								Grindley et al., 1995
		Six3								Oliver et al., 1995a
		Otx2								Mallamaci et al., 1996
		Pitx1 (Ptx1)								Lanctôt et al., 1997
		Dlx5								Yang et al., 1998
		BF1 (Foxg1)								Hatini et al., 1999
		Six4								Chen et al., 2009
			Sox2							Wood and Episkopou, 1999
			Pou2f1							Donner et al., 2007
				Nr2e1						Monaghan et al., 1995
					Msx1					Grindley et al., 1995
					Sox3					Collignon et al., 1996
					HES6					Vasiliauskas and Stern, 2000
					Eya1					Xu et al., 1997
					Eya2					Xu et al., 1997
					Eya4					Borsani et al., 1999
					Mash1					Cau et al., 1997
					Ngn1					Cau et al., 1997
					NeuroD					Cau et al., 1997
						Emx2				Simeone et al., 1992; Mallamaci et al., 1998
								Six1		Oliver et al., 1995b
								Six6		Jean et al., 1999
								Hes1		Cau et al., 2000
								Hes5		Cau et al., 2000
								Foxg1		Duggan et al., 2008
								Hey1		Leimeister et al., 1999
									Cst	Vacalla and Theil, 2002

TABLE 3: Expression of transcription factors during lens placode development.

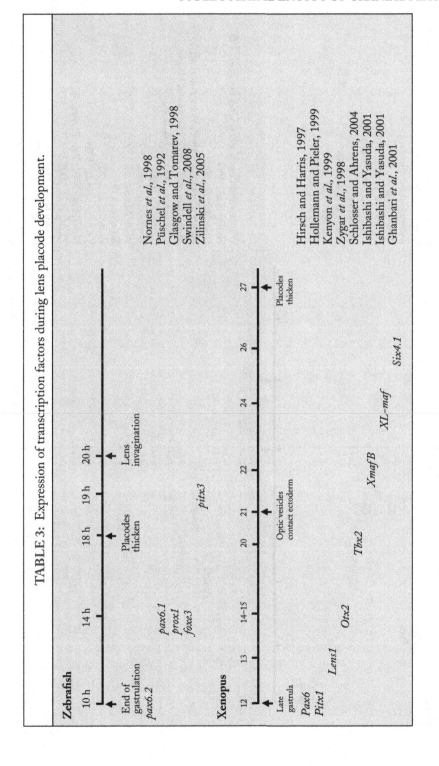

TABLE 3: (continued)

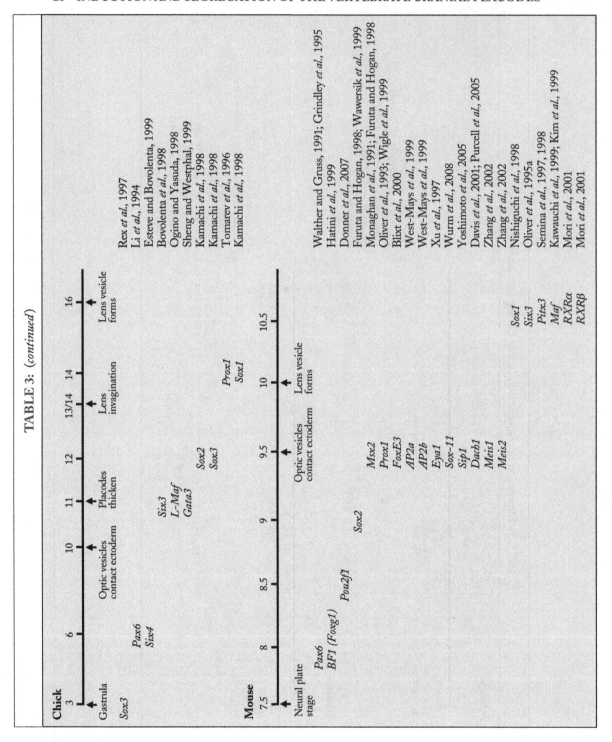

Chick

Stage	3	6	10	11	12	13/14	16	References
Event	Gastrula		Optic vesicles contact ectoderm	Placodes thicken		Lens invagination	Lens vesicle forms	
	Sox3	*Pax6*		*Six3*	*Sox2*			Rex et al., 1997
		Six4		*L-Maf*	*Sox3*			Li et al., 1994
				Gata3				Esteve and Bovolenta, 1999
								Bovolenta et al., 1998
								Ogino and Yasuda, 1998
								Sheng and Westphal, 1999
								Kamachi et al., 1998
								Kamachi et al., 1998
								Tomarev et al., 1996
								Kamachi et al., 1998

Mouse

Stage	7.5	8	8.5	9	9.5	10	10.5	References
Event	Neural plate stage				Optic vesicles contact ectoderm	Lens vesicle forms		
		Pax6	*Pou2f1*	*Sox2*	*Msx2*	*Prox1*	*Sox1*	Walther and Gruss, 1991; Grindley et al., 1995
		BF1 (Foxg1)			*Prox1*	*Sox1*	*Six3*	Hatini et al., 1999
					FoxE3		*Pitc3*	Donner et al., 2007
					AP2a		*Maf*	Furuta and Hogan, 1998; Wawersik et al., 1999
					AP2b		*RXRa*	Monaghan et al., 1991; Furuta and Hogan, 1998
					Eya1		*RXRβ*	Oliver et al., 1993; Wigle et al., 1999
					Sox-11			Blixt et al., 2000
					Sip1			West-Mays et al., 1999
					Dach1			West-Mays et al., 1999
					Meis1			Xu et al., 1997
					Meis2			Wurm et al., 2008
								Yoshimoto et al., 2005
								Davis et al., 2001; Purcell et al., 2005
								Zhang et al., 2002
								Zhang et al., 2002
								Nishiguchi et al., 1998
								Oliver et al., 1995a
								Semina et al., 1997, 1998
								Kawauchi et al., 1999; Kim et al., 1999
								Mori et al., 2001
								Mori et al., 2001

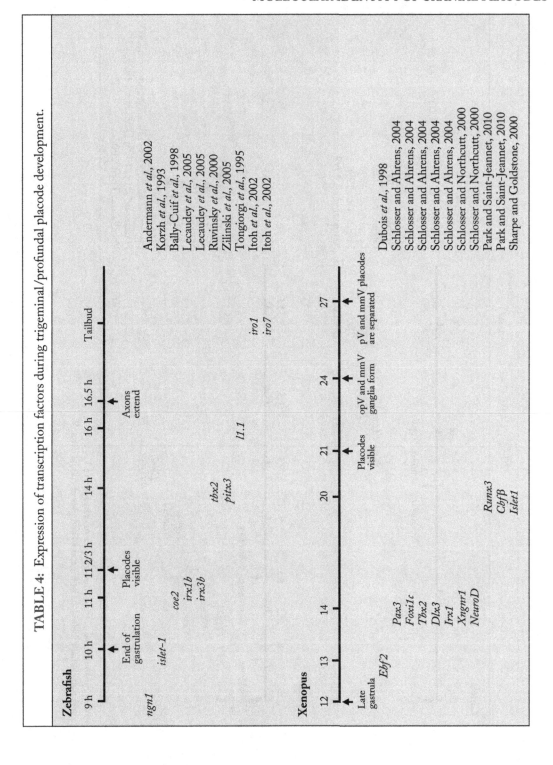

TABLE 4: Expression of transcription factors during trigeminal/profundal placode development.

TABLE 4: (*continued*)

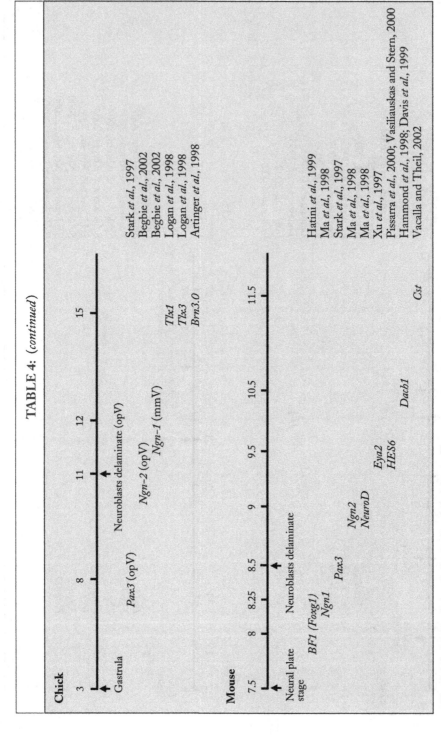

opV: ophthalmic lobe of trigeminal complex of cranial nerve V.
mmV: maxillomandibular lobe of trigeminal complex of cranial nerve V.

TABLE 5: Expression of transcription factors during otic placode development.

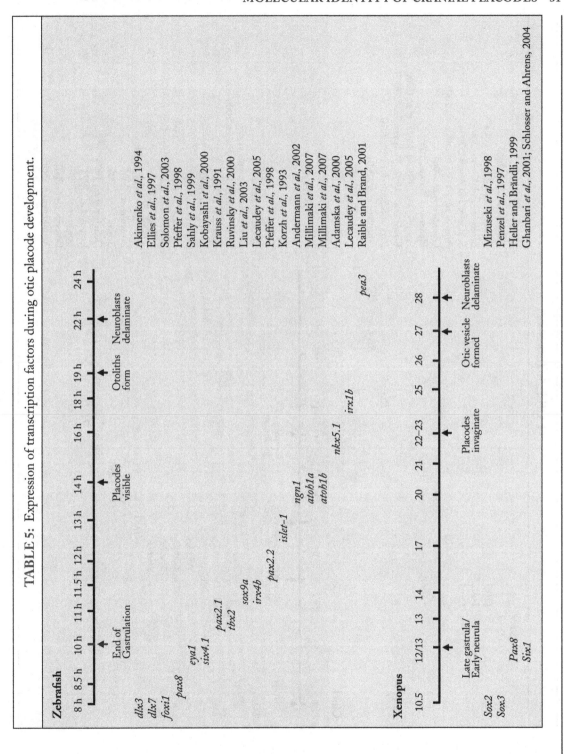

TABLE 5: (*continued*)

Chick

Timeline (stages): 3 — 6 — 8 — 9 — 10 — 11 — 12 — 13 — 14 — 15 — 17 — 21

Events along the axis: Gastrula · Placodes thicken (≈9) · Placodes invaginate (≈11) · Neuroblasts delaminate (≈14)

Genes and their approximate stage positions:

- Sox3 (Gastrula)
- Six4 (6)
- Sox9, Tbx2
- Pax2, Lmx1 (6–8)
- Dlx3, Irx1, Pax2 (8)
- Dlx3 (8–9)
- Nkx5.1, Gbx2, Spalt4, Islet-1 (9–10)
- Sox10 (10)
- Gata3 (11)
- Six2 (12)
- Dlx5 (12)
- Dlx6 (13)
- Pax2 (13–14)
- Grg 4, Grg 5 (14)
- SoHo1 (14)
- Tlx1 (15)
- Tlx3 (15)
- Gata2 (15)
- Runx1 (15)
- Six4.1 (17)
- Brn3.0 (17)
- Otx2 (21)

Citations:

Spokony et al., 2002; Saint-Germain et al., 2004
Takabatake et al., 2000
Schlosser and Ahrens, 2004
Schlosser and Ahrens, 2004
Schlosser and Ahrens, 2004
Aiko et al., 2003; Honoré et al., 2003
Ghanbari et al., 2001
Heller and Brandli, 1997
Roose et al., 1998; Molenaar et al., 2000
Roose et al., 1998; Molenaar et al., 2000
Park and Saint-Jeannet, 2010
Ghanbari et al., 2001

Rex et al., 1997
Esteve and Bovolenta, 1999
Groves and Bronner-Fraser, 2000
Giraldez, 1998
Brown et al., 2005
Herbrand et al., 1998
Hidalgo-Sánchez et al., 2000
Barembaum and Bronner-Fraser, 2007
Li et al., 2004
Sheng and Stern, 1999
Brown et al., 2005
Brown et al., 2005
Deitcher et al., 1994
Logan et al., 1998
Logan et al., 1998
Lillevälli et al., 2007
Artinger et al., 1998
Sánchez-Calderón et al., 2002

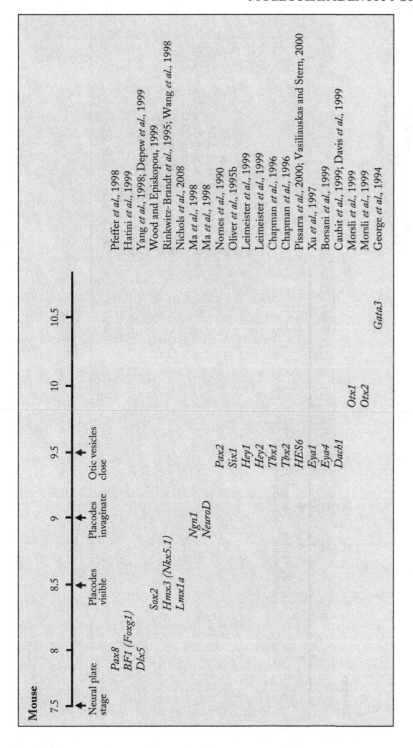

TABLE 6: Expression of transcription factors during epibranchial placode development.

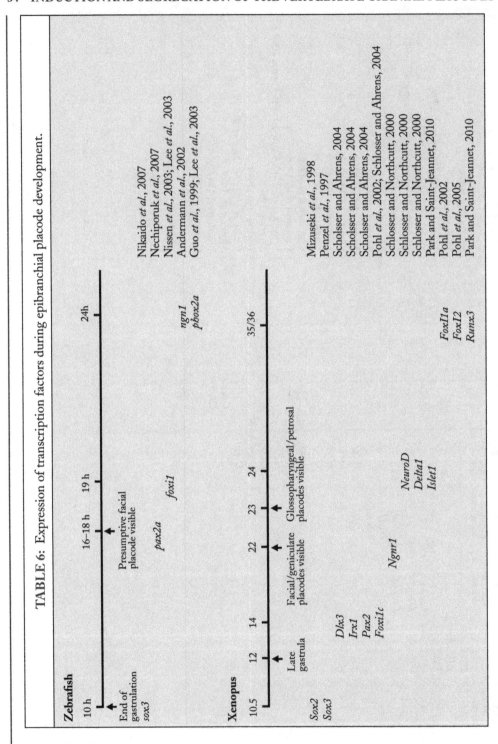

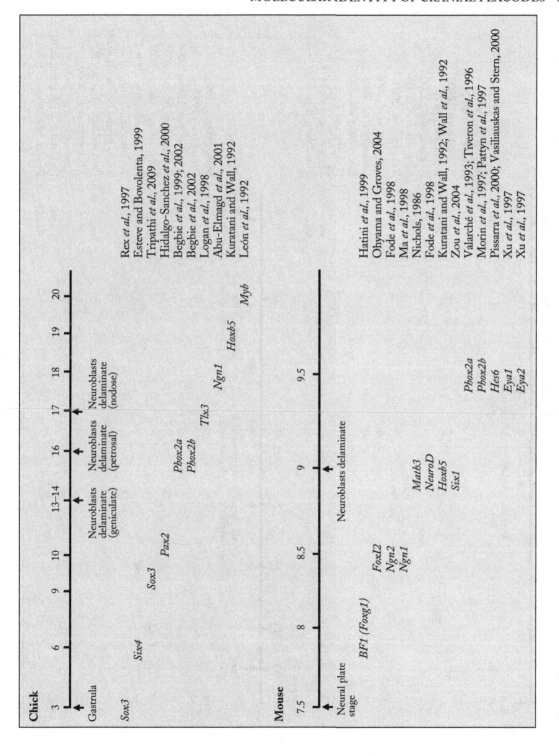

TABLE 7: Expression of transcription factors during lateral line placode development.

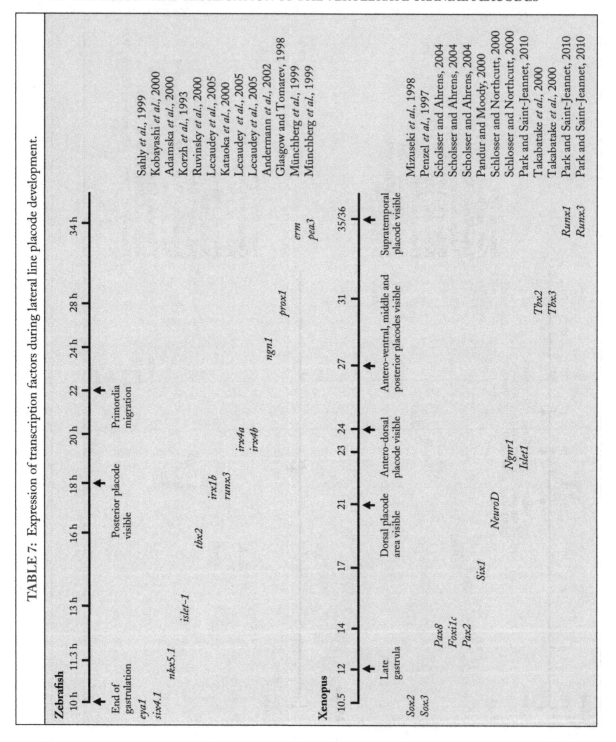

2000; Rebay et al., 2005; Kumar, 2009). *Six* and *Eya* genes were first identified in Drosophila as *sine oculis* (*so*) and *eyes absent* (*eya*). Mutations in either gene show severe eye phenotypes in flies, and Six-Eya combined overexpression can induce ectopic eyes (Bonini et al., 1993; Pignoni et al., 1997).

These genes are not only expressed in the pre-placodal ectoderm, but also later in most cranial placodes and their derivatives (see Tables 1–7 for references). Mutations in *Eya1* and *Six1* genes in humans, mice, and zebrafish show a very similar spectrum of defects in various placode derivatives suggesting that they are important regulators of placodal development. Mice hetero-zygous for *Eya1* have hearing loss due to defects of the middle ear (Xu et al., 1999), a phenotype similar to the one observed in patients affected by Branchio-Oto-Renal syndrome caused by a mutation in the *EYA1* gene (Abdelhak et al., 1997). Homozygous *Eya1* mutant mice display much more severe inner ear defects. Otic placode formation is initiated normally but development arrests at the vesicle stage and the vestibulochochlear ganglia do not form. In addition, the epibranchial placode-derived ganglia are missing (Xu et al., 1999). Mutations in the human *EYA1* gene also lead to congenital eye defects (Azuma et al., 2000). Mice lacking Six1 function display inner ear abnor-malities similar to the defects observed in *Eya1* homozygous mutant mice: the otic vesicle fails to expand leading to an absence of cochlear duct and semicircular canals (Laclef et al., 2003; Li et al., 2003; Zheng et al., 2003). Like *Eya1* mutant mice, *Six1* homozygous mutant mice lack both the vestibulochochlear and petrosal ganglia. However, a complete loss of placodal derivatives in any of these mutants has not been observed as one might expect from genes involved in establishing the pre-placodal domain. Since several members of the Six and Eya families are co-expressed in this region the lack of an early phenotype may be due to functional redundancy. Consistent with this possibility, Six1/Six4 compound mice have trigeminal ganglion defects, a phenotype not seen in the single mutants (Grifone et al., 2005; Konishi et al., 2006).

In zebrafish *dog-eared* mutants carry a mutation in the *eya1* gene. In this organism, Eya1 function appears to be primarily required for survival of sensory hair cells in the developing ear and lateral line neuromasts (Whitefield et al., 1996; Kozlowski et al., 2005). In *Xenopus* overexpression of Six1 leads to an expansion of the pre-placodal region at the expense of neural plate, neural crest, and epidermis, in contrast Six1 morpholino-mediated knockdown results in the loss of pre-placodal markers at early stage, including *Eya1* (Brugmann et al., 2004; Brugmann and Moody, 2005). A later phenotype of Six1-depleted *Xenopus* embryos has not been reported.

Altogether these observations point to an essential role for Six1 and Eya1 in early placode de-velopment. The precise molecular function of these genes in the pre-placodal region is still unclear. They could confer placodal competence to the ectoderm or bias the anterior neural plate border to-ward a placode fate as recently suggested (Streit, 2004, Bailey and Streit, 2006). It is remarkable that this network of genes has been conserved during evolution to specify sensory organs in Drosophila and in vertebrates (Schlosser, 2007).

Pax GENE FAMILY

Pax genes encode a DNA-binding domain termed the paired domain and in addition some also encode a second binding domain, a paired type homeobox. Pax genes regulate a broad array of developmental processes including proliferation, differentiation, cell adhesion, and signaling. These genes are divided into four groups based on sequence similarity: Pax1/Pax9 group, Pax2/Pax5/Pax8 group, Pax3/Pax7 group and Pax4/Pax6 group (reviewed in Dahl et al., 1997; Chi and Epstein, 2002). Among the different *Pax* genes only *Pax2/5/8*, *Pax3/7*, and *Pax6* have been implicated in placode development. All placodes express one or more Pax gene at a relatively early stage in their development. *Pax8* is an early marker of the otic placode; *Pax6* is a well-known and essential regulator of the lens placode and olfactory placodes; *Pax3* is the earliest known marker for the ophthalmic branch of the trigeminal ganglion, while *Pax2* is expressed early in both the otic and epibranchial placodes.

The three vertebrate genes of the *Pax2/5/8* group are expressed in a broad range of tissues including the kidney, thyroid, thymus and central nervous system (reviewed in Dahl et al., 1997; Chi and Epstein, 2002). Early on in embryogenesis *Pax2* and *Pax8* are also expressed in the posterior placodal area at neural plate stages, with maintained expression in the developing otic and epibranchial placodes (see Tables 1–7 for references). Work in the chick suggests that Pax2 controls epibranchial neuron identity (Baker and Bronner-Fraser, 2000). In zebrafish, Pax2 and Pax8 function synergistically to specify the otic placode (Hans et al., 2004), but act redundantly to maintain the otic placode (Mackereth et al., 2005). By contrast mice lacking Pax8 function have no obvious inner ear phenotype (Mansouri et al., 1998), while Pax2 mutant mouse embryos display agenesis of the cochlear duct and associated ganglion (Torres et al., 1996; Burton et al., 2004).

Pax3 and Pax7 are two important regulators of myogenesis, however, they are also expressed in the dorsal neural tube and during neural crest development (reviewed in Dahl et al., 1997; Chi and Epstein, 2002). Pax3 is also detected in the profundal placode where it is implicated in establishing neuron identity (Baker and Bronner-Fraser, 2000; Baker et al., 2002). In the Pax3 mouse mutant, Splotch (*Pax3^{Splotch}*), several cranial ganglia are hypoplastic including the trigeminal ganglion (Epstein et al., 1991; Tremblay et al., 1995). This phenotype is also associated with severe defects in the formation of the cochlear duct and vestibulocochlear ganglion (Buckiova and Syka, 2004). Placode defects have not been analyzed in Pax3-depleted *Xenopus* embryos (Monsoro-Burq et al., 2005).

Pax6 is implicated in the formation of the most anterior placodes (reviewed in Gehring and Ikeo, 1999; Bhattacharyya and Bronner-Fraser, 2004). Pax6 is expressed in the central nervous system and in the ectoderm that will give rise to the adenohypophyseal, olfactory and lens placodes (see Tables 1–7 for references). Pax6 is best known for regulating lens differentiation by binding directly to the enhancers of crystallin genes to regulate their expression (review in Chow and Lang, 2001; Cvekl et al., 2004). Pax6 overexpression leads to ectopic formation of eyes and lens placodes (Altmann et al., 1997; Chow et al., 1999), an activity very well conserved during evolution (Gehring and Ikeo, 1999). On the other hand Pax6 mutant mouse embryos, Small eyes (*Pax6^Sey*), show reduced eyes with missing lens and olfactory placodes (Hogan et al., 1986). In the *Pax6* mutants the adenohypophyseal placode is also defective (Bentley et al., 1999).

Pax, *Six* and *Eya* genes are part of the same regulatory network during placode development. However, *Pax* genes are expressed in a more restricted manner, suggesting that they may have a more specific role in promoting placode identity within a pre-placodal region established by *Six* and *Eya* (Baker and Bronner-Fraser, 2001; Streit, 2002, 2004; Schlosser and Ahrens, 2004; Schlosser, 2006). The mechanism by which *Pax* genes regulate placode identity is not understood, but is likely to involve others families of transcription factors (see Tables 1–7). In addition, and because their expression is maintained in derivatives of several cranial placodes Pax genes are also likely to regulate other aspects of placode development, including morphogenesis, patterning and differentiation.

Induction and Segregation of the Cranial Placodes

At the end of gastrulation, the ectoderm of the vertebrate embryo can be divided into three major domains: the non-neural ectoderm and the neural plate separated by a third region known as the neural plate border. The non-neural ectoderm and neural plate will develop into epidermis and central nervous system, respectively, while the neural plate border contains at least two cell populations: neural crest and pre-placodal ectoderm. The neural crest is located lateral to the neural plate but excluded from its most anterior region. The pre-placodal ectoderm is restricted to the anterior region of the embryo, lateral to the neural crest and at the most rostral boundary of the neural plate. The pre-placodal ectoderm abuts directly the neural plate anteriorly, and the neural crest in the lateral regions (Figure 1). The restriction of placodal fate to anterior regions of the neural plate border is not due to restrictions in competence at the gastrula stage since all regions of the ectoderm are competent to form placodes, and this competence is maintained until fairly late in development, at least for some placodes (Jacobson, 1966; Noramly and Grainger, 2002). Instead, the cranial restriction of placodal fate must be explained by local induction.

Formation of the pre-placodal region is initiated through a series of events that first define the neural plate border and subsequently subdivide the border into placode and neural crest precursors (Figure 1). This is achieved through interactions with surrounding tissues, neural plate, the future epidermis and underlying head mesoderm, all of which secrete factors controlling placode versus neural crest fate. Therefore, different signaling events converge to position the placode and the neural crest territories next to the neural plate. Because neural crest and placodes originate from the same region of the embryo, the neural plate border, it is likely that the same set of factors will be differentially deployed for the induction of both cell types (reviewed in Knecht and Bronner-Fraser, 2002; Huang and Saint-Jeannet, 2004). How these signals are integrated at the neural plate border to generate distinct fates is an important and unresolved question.

Placodes are often discussed as independent structures, as they arise from non-overlapping ectodermal thickenings. However, most current models of placode induction involve a multi-step process starting with the formation of a pre-placodal region in which individual fates are initially intermingled (Kozlowski et al., 1997; Streit, 2004; Battacharrya and Bronner-Fraser, 2004; Xu et al., 2008). Later, placodes with distinct identities segregate in a stereotypical antero-posterior pattern in response to additional inductive cues (Figure 1). This model originates primarily in the pioneering studies of Antone Jacobson performed in the 60's using amphibian embryos (Jacobson, 1963a; b; c; 1966), studies that are now largely supported by molecular analyses (reviewed in Baker and Bronner-Fraser, 2001; Streit, 2004; 2007; Battacharrya and Bronner-Fraser, 2004; Schlosser and Ahrens, 2004; Brugmann and Moody, 2005; Schlosser, 2006; Bailey and Streit, 2006; McCabe and Bronner-Fraser, 2009).

HISTORICAL PERSPECTIVE

In a series of experiments performed in two different species of amphibians, *Taricha torosa* and *Ambystoma punctatum*, Jacobson evaluated the abilities of different tissues to induce lens, otic or olfactory placodes during development (Jacobson 1966; 1963a; 1963b). These studies demonstrated that the lens and olfactory placodes can be induced by early signals from the endoderm and mesoderm, while the otic placode is induced by mesoderm- and neuroectoderm-derived signals. These signals are however not sufficient to elicit development of a complete lens, inner ear or olfactory structure. By evaluating the proximity of the presumptive placodal ectoderm to these inducers during development of the lens, Jacobson proposed that the presumptive lens ectoderm is initially induced by the endoderm, followed by inducing signals from the intervening mesoderm, and eventually from signals derived from the optic cup, adjacent to the lens placode. Based on these observations Jacobson proposed that a similar sequence of inductions could be applied to the formation of other placodes.

In another set of experiments Jacobson demonstrated that within the ectoderm adjacent to the anterior neural plate, cells are competent to give rise to any placode, but progressively lose this ability overtime, as placodal domains start to segregate within the pre-placodal ectoderm (Jacobson, 1963c). In these experiments the pre-placodal ectoderm is dissected and rotated along its antero-posterior axis at the early neurula stage, so that prospective ear ectoderm end up next to forebrain, and prospective nose ectoderm next to hindbrain (Figure 10A). Within the transplanted pre-placodal ectoderm, placodes develop according to their novel (ectopic) position along the antero-posterior axis, with occasional formation of anterior ectopic ears (Figure 10B). In the same experiments performed at a later stage in development, placodes developed according to their original position, suggesting that the identity of the pre-placodal ectoderm was already determined at the stage of transplantation. However, some plasticity persists in the transplanted ectoderm since olfactory tissues are induced adjacent to the ectopic anterior ears, and ears are induced next to ectopic posterior noses (Figure 10B). Therefore, at early neurula stage cells within the pre-placodal ectoderm are competent to generate a placode distinct from their normal fate, while at late neurula stage their fate ap-

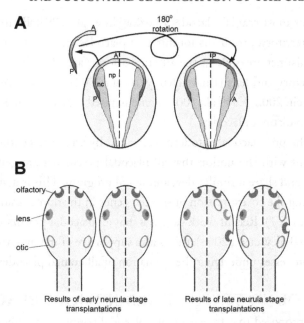

FIGURE 10: Diagrams illustrating Antone Jacobson's pre-placodal ectoderm rotation experiments. (A) The pre-placodal ectoderm of an amphibian neurula stage donor embryo was dissected and transplanted on a host embryo so that its antero-posterior polarity was reversed (180° rotation). (B) The outcome of these experiments is shown in the context of early (left panels) and late (right panels) neurula stage transplantations showing differences in the level of determination of the pre-placodal ectoderm. In each case the manipulated half is on the right side. Modified from Jacobson (1963c).

pears to be determined (Jacobson, 1963c). These experiments highlight three key points regarding the specification of cranial placodes: (i) all placodes originate from a common domain; (ii) there is a shared inductive mechanism for all placodes; and (iii) placode induction is a multi-step process.

A COMMON DOMAIN FOR ALL CRANIAL PLACODES

The pre-placodal ectoderm surrounding the anterior neural plate at the late gastrula/early neurula stages has been described in mouse, chick, amphibian and fish embryos. A large body of work, including that of Antone Jacobson described above, has shown that induction of different placodes involves similar sequential interactions between the presumptive pre-placodal region and surrounding tissues (reviewed in Baker and Bronner-Fraser, 2001).

Lineage and fate map analyses in fish and chick embryos at the gastrula and neurula stage have demonstrated that this region of the embryonic ectoderm, adjacent to the prospective brain contains precursors for most cranial placodes (Kozlowski et al., 1997; Bhattacharyya et al., 2004; Streit, 2002). In this territory, precursors for different placodes are initially intermingled and eventually recruited into distinct areas along the antero-posterior axis. Using focal dye labeling, recent fate maps of the olfactory and lens (Bhattacharyya et al., 2004), and the trigeminal, epibranchial, and otic placodes (Streit, 2002; Xu et al., 2008) demonstrated that adjacent ectodermal cell populations can contribute to distinct placodes.

The fact that the pre-placodal ectoderm is formed by an heterogenous population placodal precursors is consistent with the notion that all placodal precursors are generated by a common inductive mechanism and share a similar developmental program. This is a view supported by molecular analyses showing the co-expression of specific genes in the domain surrounding the anterior neural plate (see Tables 1–7). Recent work suggests that all placodal precursors start with the same initial fate, a lens fate (Bailey, et al., 2006); this is also supportive of the hypothesis of a commonality of organization and processes underlying the formation of all cranial placodes.

LENS, THE GROUND STATE OF ALL SENSORY PLACODES

Recently, it has been proposed that the entire pre-placodal region is initially specified as lens (Bailey, et al., 2006), the simplest form of all ectodermal placodes, giving rise to only two cell types (Jacobson, 1966). When antero-posterior segments of the pre-placodal ectoderm are cultured in isolation in a defined medium, all segments form lens structures and express lens-specific genes, even cells that normally never contribute to the lens, like the most posterior segmentsof pre-placodal ectoderm. Moreover, none of these explants express markers specific for olfactory, trigeminal or otic placodes. All placodal cells, regardless of their final fate have an initial lens character, which represents the ground state for all sensory placodes (Bailey, et al., 2006). This model implies that in normal placode development the acquisition of a specific identity requires an initial step of repression of lens character, prior to or concomitant to the activation of a program specific for olfactory, trigeminal or otic fate. Since lens placode is non-neurogenic, non-lens placodes also need to acquire neurogenic properties in the process.

Among the possible candidate molecules repressing lens character in the pre-placodal region are members of Fibroblast Growth Factors (Fgf) family. As recently described, exposure of the presumptive lens ectoderm to Fgf8 blocks expression of the lens marker *Pax6* and promotes olfactory placode character. On the other hand, markers for adenohypophyseal, trigeminal or otic placodes are not induced under these conditions (Bailey, *et al.*, 2006). These results indicate that Fgf8 signaling represses lens fate and is sufficient to elicit olfactory placode development. This work also suggests that a different signal (or combination of signals) might be required to specify other placodes. Alternatively, because several members of the Fgf family (including Fgf3, Fgf10 and Fgf19) are implicated at least in adenohypophyseal, otic and epibranchial placodes formation in several species (see below), other Fgf ligands besides Fgf8, might be implicated in the specification of other placodes.

INDUCING FACTORS IN THE GENERATION OF CRANIAL PLACODES

The initial overlap between different placode precursors within the pre-placodal ectoderm has made it very challenging to identify the specific inductive signals for the different placodes. Several classes of signaling molecules are implicated in the induction of the pre-placodal region and its subsequent sub-division into domains with distinct placodal identities. A number of recent review articles have discussed some of these findings (Baker and Bronner-Fraser, 2001; Bailey and Streit, 2006; Schlosser, 2006; McCabe and Bronner-Fraser, 2009); here we will summarize the activity of these molecules focusing more specifically on the Bone Morphogenetic Protein (Bmp), Fibroblast Growth Factor (Fgf) and Wnt signaling pathways.

Bone Morphogenetic Protein Signaling

While unperturbed Bmp signaling in the ectoderm converts cells into epidermis, attenuation of Bmp activity in the ectoderm results in emergence of neural, neural crest and placodal cells (for review Sasai and De Robertis, 1997; Stern, 2005). Studies in frog and fish indicate the neural plate border forms in region of the ectoderm where Bmp signaling is partially attenuated by Bmp antagonists, such as Chordin, Noggin and Follistatin, derived from the axial mesoderm (Marchant et al., 1998; Nguyen et al., 1998; Tribulo et al., 2003). In *Xenopus* a balance of Bmps and their antagonists is in part responsible for positioning the pre-placodal ectoderm (Brugmann et al., 2004; Glavic et al., 2004). *Xenopus* animal caps treated with different concentrations of the Bmp antagonist, Noggin, form epidermis in the presence of high levels of Bmp activity, while neural crest and pre-placodal cells are generated at intermediate levels and neural plate at low levels (Wilson, et al., 1997; Tribulo, et al., 2003; Brugmann, et al., 2004; Glavic, et al., 2004). Zebrafish embryos with mutations in distinct components of the Bmp signaling pathway show expanded neural crest domain while the placode territory is somewhat displaced, but not expanded (Neave, et al., 1997; Nguyen, et al., 1998).

Bmp signals are also implicated in the specification of olfactory and lens placodes in the chick embryo. At gastrula stages Bmp2 and Bmp4 are expressed in the pre-placodal region. In an explant assay using placodal progenitor cells, short exposure to Bmp signaling promotes specification of olfactory fate, while prolonged exposure to Bmp signals promotes formation of lens cells at the expense of placodal cells. Bmp signaling is also sufficient to promote olfactory and lens progenitors in forebrain explants (Sjodal et al., 2007).

Bmp4-/- mouse embryos lack lens placodes, but expression of Six3 and Pax6 is detected in the prospective lens ectoderm and the olfactory placode appears normal (Furuta and Hogan, 1998).

These results suggest that placodal progenitor cells are induced without Bmp4, which may reflect functional redundancy with other Bmp family members such as Bmp7. These results also indicate that Bmp4 activity is required for differentiation of lens but not olfactory placodal cells after the initial specification of placodal progenitors. Bmp4 alone is not sufficient to promote lens development suggesting that it may act with other inducers (Furuta and Hogan, 1998). Bmp7 protein is present in the head ectoderm at the time of lens placode induction. Inhibition of Bmp7 signaling at the time of lens placode induction significantly decreases the frequency of lens formation in an organ culture system. The expression of the lens placode marker Sox2 was also severely affected in *Bmp7-/-* mutant embryos (Wawersik, 1999).

Fibroblast Growth Factor Signaling

In the chick, several observations implicate Fgfs as one of the key factors initiating the formation of the border region: misexpression of Fgf8 induces ectopic expression of neural plate border-specific genes (Streit and Stern, 1999; Litsiou et al., 2005). However, in chich and frogs Fgf alone is not sufficient to generate neural crest and placode fates (Mayor, *et al.,* 1997; LaBonne and Bronner-Fraser, 1998; Monsoro-Burq, *et al.,* 2003; Ahrens and Schlosser, 2005; Litsiou, *et al.,* 2005; Hong et al., 2008). There is good evidence that Fgf8 may be required in concert with Bmp signaling to promote placodal fate. In *Xenopus,* Fgf8 positively regulates expression of the pre-placodal specific gene *Six1* when Bmp is inhibited in the ectoderm (Ahrens and Schlosser, 2005). Fgf8 is expressed in the paraxial mesoderm and anterior neural ridge in frogs (Christen and Slack, 1997), and morpholino-mediated knockdown of Fgf8a results in a broad loss of neural crest and pre-placodal genes (Hong and Saint-Jeannet, 2007). However, the loss neural crest in these experiments is likely to be indirect since Fgf8a is required to activate Wnt8 in the mesoderm (Hong et al., 2008). As described above, in chick embryos Fgf8 is also hypothesized to be the factor required to repress lens fate in the pre-placodal ectoderm and to promote specification of olfactory cells (Bailey et al., 2006).

Several members of the Fgf family are implicated in otic placode specification (reviewed in Schimmang 2007; Schneider-Maunoury and Pujades, 2007). Different species appear to use different combinations of these molecules. Fgf19 expressed in the paraxial mesoderm and Fgf8 derived from the endoderm, are implicated in otic placode induction in the chick (Ladher et al., 2000; Ladher et al., 2005). In zebrafish otic placode induction depends primarily on Fgf3 and Fgf8 (Philipps et al., 2001; Maroon et al., 2002; Leger and Brand, 2002; Liu et al., 2003). In the mouse, paraxial mesoderm-derived Fgf8 and Fgf10 are involved in this process (Wright and Mansour, 2003; Ladher et al., 2005; Zelarayan et al., 2007).

A role for Fgf3 and Fgf8 ligands is reported in epibranchial placode specification in zebrafish (Nechiporuk, et al., 2007; Nikaido, et al., 2007; Sun, et al., 2007). In the Fgf8 mutant embryos, *acerebellar (ace)*, placodal expression of Sox3 was disrupted. This phenotype was rescued by implantation of an Fgf8 bead near the prospective hindbrain. This requirement for Fgf signaling was further demonstrated using a soluble FgfR inhibitor, which resulted in reduced expression of the epibranchial markers Sox3 and Phox2a (Nikaido, et al., 2007). In contrast, other experiments using morpholinos against both Fgf3 and Fgf8 suggested a dual requirement of Fgf3 and Fgf8, rather than Fgf8 alone (Sun et al., 2007; Nechiporuk et al., 2007).

Fgf3 is prodiced by the ventral diencephalon and is required for the expression of early adenohypophysis markers (Herzog et al., 2004). In the mouse, loss of FgfR2b or deletion of its ligand, Fgf10, leads to early defects in the adenohypophysis (Ohuchi et al., 2000) also suggesting an important function of Fgf signaling in adenohypohysis placode formation.

Wnt Signaling

At the neural plate border, canonical Wnt signaling is required in conjunction with Bmp attenuation to specify the neural crest (reviewed in Knecht and Bronner-Fraser, 2002; Huang and Saint-Jeannet, 2004). Interestingly, the pre-placodal ectoderm appears to have a different requirement with regard to Wnt signaling compared to the neural crest, because inhibition of the canonical Wnt signaling pathway favors placodal tissue at the expense of neural crest fate (Brugmann et al., 2004; Litsiou et al., 2005). Recent work using *Xenopus* animal caps suggest that inhibition of canonical Wnt signaling is more specifically required to define the most anterior domain (olfactory) of the pre-placodal ectoderm (Park and Saint-Jeannet, 2008). Expression of Fgf8a promotes olfactory placode formation (based on Dmrt4 expression) in animal caps injected with Noggin. However, simultaneous activation of canonical Wnt signaling in these explants inhibited Dmrt4 expression and promoted otic placode fate based on Pax8 expression. These results indicate that cranial placodes along the antero-posterior axis have different requirements with regard to Wnt signaling. The most anterior cranial placode (olfactory, at least) requires Wnt inhibition while more posterior placodes (otic) depend on active canonical Wnt signaling (Park and Saint-Jeannet, 2008).

There is also strong evidence in multiple species that canonical Wnt signaling is involved in otic placode induction. In chick, presumptive otic ectoderm showed a stronger induction of the otic marker gene *Pax2* when cultured in the presence of Fgf19 and Wnt8C compared to explants cultured with Fgf19 alone (Ladher et al., 2000). A subsequent study in zebrafish showed that Wnt8 depletion or overexpression of a Wnt inhibitor (Dkk1) did not prevent otic vesicle formation (Phillips et al., 2004). Mouse work suggests that the presumptive otic ectoderm is exposed to Wnt signals early, as the activity of a TCF/Lef-LacZ reporter is detected in the pre-otic ectoderm (Ohyama et al., 2006). Conditional knockout of β-catenin results in a smaller than normal otic vesicle, and conditional stabilization of β-catenin expands the placodal domain. In the mouse canonical Wnt signaling is likely to mediate the decision between otic and epidermal fate (Ohyama et al., 2006;

2007). In *Xenopus*, canonical Wnt signaling appear to cooperate with Fgf signals (Fgf3 and Fgf8) to specify the otic placode (Park and Saint-Jeannet, 2008).

Wnt molecules are also implicated in trigeminal placode formation in the chick embryo. Blocking canonical Wnt signaling prevented the targeted cells to adopt or maintain an ophthalmic trigeminal placodal fate, based on the expression of Pax3 and Eya4. In contrast, activation of the Wnt pathway was not sufficient to elicit Pax3 expression, suggesting that other signaling cues are also required to promote ophthalmic trigeminal placodal fate (Lassiter et al., 2007; Dude et al., 2009).

Other Signaling Pathways

In addition to the three major signaling pathways discussed above other signaling molecules are also implicated in various aspects of placode development. While very little is known about the factors inducing lateral line placodes in amphibia and fish, recent studies are focuse on the signaling molecules regulating posterior lateral line primordium migration in zebrafish. Chemokines of the CXCL class and their receptors (CXCR) are known for promoting the directional migration of leukocytes during inflammation (Zlotnik and Yoshie, 2000). In zebrafish, this signaling pathway is essential to provide directionality during migration of the posterior lateral line primordium. Moreover the cooperation between chemokine, Wnt and Fgf signaling appears to regulate the mechanisms by which the primordium maintains its integrity during migration despite the periodic deposition of cells during formation of the lateral line system (Dambly-Chaudiere et al., 2007; Aman and Piotrowski, 2008; Nechiporuk and Raible, 2008). The same class of molecules is also implicated in olfactory placode development. In zebrafish, Cxcl12/Cxcr4 signaling mediates assembly of olfactory placodal precursors into a compact cluster to form the olfactory placode (Miyasaka et al., 2007). Later in development, Cxcr4-mediated chemokine signaling is required for assembling the trigeminal sensory neurons into a ganglion (Knaut et al., 2005).

Platelet Derived Growth Factor (Pdgf) signaling is implicated in the induction of the ophthalmic lobe of the trigeminal placode in the chicken (McCabe and Bronner-Fraser, 2008). Pdgf receptor β is detected in the cranial ectoderm at the time of trigeminal placode formation, and the ligand Pdgfd is expressed in neural folds of the midbrain. In recombinants explants of quail ectoderm with chick neural tube, which normaly promote trigeminal placode fate, blocking Pdgf signaling results in loss of Pax3 and CD151 expression, two early markers of the trigeminal placode. Conversely, microinjection of exogenous Pdgfd increases the number of Pax3-expressing cells in the trigeminal placode and neurons in the condensing ganglia.

Shh is one of the primary factors involved in adenophypophysis placode induction. During development Shh is expressed throughout the oral ectoderm, but it is excluded from Rathke's pouch as soon as this structure forms. In Shh-deficient mouse embryos, formation of the diencephalon is severely disrupted which has made it difficult to assess adenohypophysis development. However, Rathke's pouch formation was completely arrested in transgenic animals expressing a specific hedgehog inhibitor (Hip) throughout the oral ectoderm (Treier et al., 2001). In the talpid[3] chicken mutant Shh signaling is reduced, and formation of the pituitary is severely disrupted (Lewis et al., 1999). Large ectopic lenses form as diverticula from the roof of the mouth, at a position corresponding to the adenohypophyseal placode in normal embryos (Ede and Kelley, 1964). The zebrafish double mutants for *shh* and *tiggy winkle hedgehog* (*twhh*), which encode partially redundant hedgehog ligands, have a complete loss of anterior pituitary fates (Herzog et al., 2003). In contrast overexpression of *shh* causes induction of an excessive number of pituitary cells at the expense of lens precursors (Dutta et al., 2005; Herzog et al., 2003; Sbrogna et al., 2003; reviewed in Pogoda and Hammerschmidt, 2009)

Retinoic acid (RA) is primarily implicated in morphogenesis and patterning of the otocyst. Its activity is presumably indirect mediated through its ability to regulate several target genes in the hindbrain, among which are Hox genes that provide rhombomere identity in the hindbrain (reviewed in Romand et al., 2006). In zebrafish, application of a dose of retinoic acid that does not perturb patterning of the anterior neural plate leads to increased otic induction, a process dependent on Fgf signaling (Hans et al., 2007). In vitamin A-deficient quail embryos, Rathke's pouch fails to develop, suggesting an important role of retinoic acid in adenohypophyseal placode development in birds. However, this phenotype might be secondary to the loss of other signaling molecules since Bmp2, Shh and Fgf8 are downregulated in these animals (Maden et al., 2007).

Conclusion

The past 10 years has seen a resurgence of interest for the study of the induction and development of vertebrate cranial placodes. This has led to significant progress facilitated by the characterization of a large repertoire of genes expressed at various stages of placode development. It is now well accepted that all cranial placodes arise initially from a common pre-placodal region, which signify that all cranial placode may have a common developmental history, and may employ similar developmental mechanisms. This is exemplified by the finding that cranial placodes initially share the same ground state, a lens fate, which represents the simplest form of all placodes. Later in development and under the influence of local signals, placode precursors become progressively divided into a number of primordia with different identities from which individual placodes will develop and differentiate. An important area of investigation for the future is to determine the exact nature and sequence of the inductive signals involved in placode specification and diversification. The information available on the repertoire of transcription factors expressed in the pre-placodal ectoderm and its derivatives has been growing very rapidly in the last few years. However, future work using high throughput screens will be needed to uncover additional players in this regulatory network in order to fully understand the mechanisms by which they act to regulate placode development.

Acknowledgments

We thank Trish Labosky for comments on the manuscript. We thank Dr. Young-Hoon Lee for help in the preparation of Figure 4. Work in J.-P. S.-J.'s lab is supported by a grant from the National Institutes of Health (RO1-DC07175).

References

Abdelhak, S., Kalatzis, V., Heilig, R., Compain, S., Samson, D., Vincent, C., Weil, D., Cruaud, C., Sahly, I., Leibovici, M., Bitner-Glindzicz, M., Francis, M., Lacombe, D., Vigneron, J., Charachon, R., Boven, K., Bedbeder, P., Van Regemorter, N., Weissenbach, J., and Petit, C. (1997). A human homologue of the Drosophila eyes absent gene underlies Branchio–Oto–Renal (BOR) syndrome and identifies a novel gene family. *Nat Genet* **15**, pp. 157–164.

Abu-Elmagd, M., Ishii, Y., Cheung, M., Rex, M., Le Rouëdec, D., and Scotting, P. J. (2001). cSox3 expression and neurogenesis in the epibranchial placodes. *Dev Biol* **237**, pp. 258–269.

Adamska, M., Léger, S., Brand, M., Hadrys, T., Braun, T., and Bober, E. (2000). Inner ear and lateral line expression of a zebrafish *Nkx5–1* gene and its downregulation in the ears of FGF8 mutant, *ace Mech Dev* **97**, pp. 161–165.

Ahrens, K., and Schlosser, G. (2005). Tissues and signals involved in the induction of placodal Six1 expression in *Xenopus laevis Dev Biol* **288**, pp. 40–59.

Akimenko, M. A., Ekker, M., Wegner, J., Lin, W., and Westerfield, M. (1994). Combinatorial expression of three zebrafish genes related to *Distal-less*: part of a homeobox gene code for the head. *J Neurosci* **14**, pp. 3475–3486.

Altmann, C. R., Chow, R. L., Lang, R. A., and Hemmati-Brivanlou, A. (1997). Lens induction by Pax-6 in Xenopus laevis. *Dev Biol* **185**, pp.119–123.

Aman, A., and Piotrowski, T. (2008). Wnt/beta-catenin and FGF signaling control collective cell migration by restricting chemokine receptor expression. *Dev Cell* **15**, pp. 749–761.

Aman, A., and Piotrowski, T. (2009). Multiple signaling interactions coordinate collective cell migration of the posterior lateral line primordium. *Cell Adh & Migr* **3**, pp. 365–368.

Andermann, P., Ungos, J., and Raible, D. W. (2002). Neurogenin1 defines zebrafish cranial sensory ganglia precursors. *Dev Biol* **251**, pp. 45–58.

Aoki, Y., Saint-Germain, N., Gyda, M., Magner-Fink, E. K., Lee, Y-H., Credidio, C., and Saint-Jeannet, J-P. (2003). Sox10 regulates the development of the neural crest-derived melanocytes in Xenopus. *Dev Biol* **259**, pp. 19–33.

Artinger, K. B., Fedtsova, N., Rhee, J. M., Bronner-Fraser, M., and Turner, E. (1998). Placodal origin of Brn-3-expressing cranial sensory neurons. *J Neurobiol* **36**, pp. 572–585.

Asa, S. L., and Ezzat, S. (2004). Molecular basis of pituitary development and cytogenesis. *Front Horm Res* **32**, pp. 1–19.

Ayer-Le Lievre, C. S., and Le Douarin, N. M. (1982). The early development of cranial sensory ganglia and the potentialities of their component cells studied in quail-chick chimeras. *Dev Biol* **94**, pp. 291–310.

Azuma, N., Hirakiyama, A., Inoue, T., Asaka, A., and Yamada, M. (2000). Mutations of a human homologue of the Drosophila eyes absent gene (EYA1) detected in patients with congenital cataracts and ocular anterior segment anomalies. *Hum Mol Genet* **9**, pp. 363–366.

Bailey, A. P., Bhattacharyya, S., Bronner-Fraser, M., and Streit, A. (2006). Lens specification is the ground state of all sensory placodes, from which FGF promotes olfactory identity. *Dev Cell* **11**, pp. 505–517.

Bailey, A. P., and Streit, A. (2006). Sensory organs: making and breaking the pre-placodal region. *Curr Top Dev Biol* **72**, pp. 167–204.

Baker, C. V. H., and Bronner-Fraser, M. (2000). Establishing neuronal identity in vertebrate neurogenic placodes. *Development* **127**, pp. 3045–3056.

Baker, C. V., and Bronner-Fraser, M. (2001). Vertebrate cranial placodes I. Embryonic induction. *Dev Biol* **232**, pp. 1–61.

Baker, C. V., Stark, M. R., and Bronner-Fraser, M. (2002). Pax3-expressing trigeminal placode cells can localize to trunk neural crest sites but are committed to a cutaneous sensory neuron fate. *Dev Biol* **249**, pp. 219–236.

Baker, C. V., Stark, M. R., Marcelle, C., and Bronner-Fraser, M. (1999). Competence, specification and induction of Pax-3 in the trigeminal placode. *Development* **126**, pp. 147–156.

Baker, C. V. H., O'Neill, P., and McCole, R. B. (2008). Lateral line, otic and epibranchial placodes: developmental and evolutionary links? *J Exp Zool (Mol Dev Evol)* **310B**, pp. 370–383.

Bally-Cuif, L., Dubois, L., and Vincent, A. (1998). Molecular cloning of Zcoe2, the zebrafish homolog of *Xenopus Xcoe2* and mouse *EBF-2*, and its expression during primary neurogenesis. *Mech Dev* **77**, pp. 85–90.

Barald, K. F., and Kelley, W. W. (2004). From placode to polarization: new tunes in inner ear development. *Development* **131**, pp. 4119–4130.

Barembaum, M., and Bronner-Fraser, M. (2007). Spalt4 mediates invagination and otic placode gene expression in cranial ectoderm. *Development* **134**, pp. 3805–3814.

Barth, K. A., and Wilson, S. W. (1995). Expression of zebrafish *nk2.2* is influenced by *sonic hedgehog/vertebrate hedgehog-1* and demarcates a zone of neuronal differentiation in the embryonic forebrain. *Development* **121**, pp. 1755–1768.

Bhattacharyya, S., and Bronner-Fraser, M. (2004). Hierarchy of regulatory events in sensory placode development. *Curr Opin Genet Dev* **14**, pp. 520–526.

Bhattacharyya, S., Bailey, A. P., Bronner-Fraser, M., and Streit, A. (2004). Segregation of lens and olfactory precursors from a common territory: cell sorting and reciprocity of Dlx5 and Pax6 expression. *Dev Biol* **271**, pp. 403–414.

Begbie, J., and Graham, A. (2001). The ectodermal placodes: a dysfunctional family. *Philos Trans R Soc London (Ser: B Biol Sci)* **356**, pp. 1655–1660.

Begbie, J., Brunet, J.-F., Rubenstein, J. L. R., and Graham, A. (1999). Induction of the epibranchial placodes. *Development* **126**, pp. 895–902.

Begbie, J., Ballivet, M., and Graham, A. (2002). Early steps in the production of sensory neurons by the neurogenic placodes. *Mol Cell Neurosci* **21**, pp. 502–511.

Bentley, C. A., Zidehsarai, M. P., Grindley, J. C., Parlow, A. F., Barth-Hall, S., and Roberts, V. J. (1999). Pax6 is implicated in murine pituitary endocrine function. *Endocrine* **10**, pp. 171–177.

Bessarab, D. A., Chong, S. W., and Korzh, V. (2004). Expression of zebrafish six1 during sensory organ development and myogenesis. *Dev Dyn* **230**, pp. 781–786.

Blixt, A., Mahlapuu, M., Aitola, M., Pelto-Huikko, M., Enerback, S., and Carlsson, P. (2000). A forkhead gene, FoxE3, is essential for lens epithelial proliferation and closure of the lens vesicle. *Genes Dev* **14**, pp. 245–254.

Borsani, G., DeGrandi, A., Ballabio, A., Bulfone, A., Bernard, L., Banfi, S., Gattuso, C., Mariani, M., Dixon, M., Donnai, D., Metcalfe, K., Winter, R., Robertson, M., Axton, R., Brown, A., van Heyningen, V., and Hanson, I. (1999). EYA4, a novel vertebrate gene related to *Drosophila eyes absent*. *Hum Mol Genet* **8**, pp. 11–23.

Bonini, N. M., Leiserson, W. M., and Benzer, S. (1993). The eyes absent gene: genetic control of cell survival and differentiation in the developing Drosophila eye. *Cell* **72**, pp. 379–395.

Bovolenta, P., Mallamaci, A., Puelles, L., and Boncinelli, E. (1998). Expression pattern of *cSix3*, a member of the Six/sine oculis family of transcription factors. *Mech Dev* **70**, pp. 201–203.

Brown, S. T., Wang, J., and Groves, A. K. (2005). Dlx gene expression during chick inner ear development. *J Comp Neurol* **483**, pp. 48–65.

Brugmann, S. A., and Moody, S. A. (2005). Induction and specification of the vertebrate ectodermal placodes: precursors of the cranial sensory organs. *Biol Cell* **97**, pp. 303–319.

Brugmann, S. A., Pandur, P. D., Kenyon, K. L., Pignoni, F., and Moody, S. A. (2004). Six1 promotes a placodal fate within the lateral neurogenic ectoderm by functioning as both a transcriptional activator and repressor. *Development* **131**, pp. 5871–5881.

Buck, L. B. (2000). The molecular architecture of odor and pheromone sensing in mammals. *Cell* **100**, pp. 611–618.

Buckiova, D., and Syka, J. (2004). Development of the inner ear in Splotch mutant mice. *Neuroreport* **125**, pp. 2001–2005.

Burgess, R., Lunyak, V., and Rosenfeld, M. (2002). Signaling and transcriptional control of pituitary development. *Curr Opin Genet Dev* **12**, pp. 534–539.

Burd, G. D. (1999). Development of the olfactory systems in the African clawed frog, *Xenopus laevis. The Biology of Early influences* (ed. R. L. Hyson and F. Johnson) pp. 153–170. New York: Kluwer Academic/Plenum Publishers.

Burns, C. J., and Vetter, M. L. (2002). Xath5 regulates neurogenesis in the Xenopus olfactory placode. *Dev Dyn* **225**, pp. 536–543.

Burton, Q., Cole, L. K., Mulheisen, M., Chang, W., and Wu, D. K. (2004). The role of Pax2 in mouse inner ear development. *Dev Biol* **272**, pp. 161–175.

Cau, E., Gradwohl, G., Fode, C., and Guillemot, F. (1997). Mash1 activates a cascade of bHLH regulators in olfactory neuron progenitors. *Development* **124**, pp. 1611–1621.

Cau, E., Gradwohl, G., Casarosa, S., Kageyama, R., and Guillemot, F. (2000). Hes genes regulate sequential stages of neurogenesis in the olfactory epithelium. *Development* **127**, pp. 2323–2332.

Caubit, X., Thangarajah, R., Theil, T., Wirth, J., Nothwang, H. G., Ruther, U., and Krauss, S. (1999). Mouse Dac, a novel nuclear factor with homology to *Drosophila* dachshund shows a dynamic expression in the neural crest, the eye, the neocortex, and the limb bud. *Dev Dyn* **214**, pp. 66–80.

Chapman, D. L., Garvey, N., Hancock, S., Alexiou, M., Agulnik, S. I., Gibson-Brown, J. J., Cebra-Thomas, J., Bollag, R. J., Silver, L. M., and Papaioannou, V. E. (1996). Expression of the T-box family genes, *Tbx1-Tbx5*, during early mouse development. *Dev Dyn* **206**, pp. 379–390.

Chen, B., Kim, E. H., and Xu, P. X. (2009). Initiation of olfactory placode development and neurogenesis is blocked in mice lacking both Six1 and Six4. *Dev Biol* **326**, pp. 75–85.

Chi, N., and Epstein, J. A. (2002). Getting your Pax straight: Pax proteins in development and disease. *Trends Genet* **18**, pp. 41–47.

Chow, R. L., and Lang, R. A. (2001). Early eye development in vertebrates. *Annu Rev Cell Dev Biol* **17**, pp. 255–296.

Chow, R. L., Altmann, C. R., Lang, R. A., and Hemmati-Brivanlou, A. (1999). Pax6 induces ectopic eyes in a vertebrate. *Development* **126**, pp. 4213–4222.

Christen, B., and Slack, J. M. W. (1997). FGF-8 is associated with anteroposterior patterning and limb regeneration in Xenopus. *Dev Biol* **192**, 455–466.

Cobos, I., Shimamura, K., Rubenstein, J. L., Martinez, S., and Puelles, L. (2001). Fate map of the avian anterior forebrain at the four-somite stage, based on the analysis of quail-chick chimeras. *Dev Biol* **239**, pp. 46–67.

Collignon, J., Sockanathan, S., Hacker, A., Cohen-Tannoudji, M., Norris, D., Rastan, S., Steva-novic, M., Goodfellow, P. N., and Lovell-Badge, R. (1996). A comparison of the properties of *Sox-3* with *Sry* and two related genes *Sox-1* and *Sox-2*. *Development* **122**, pp. 509–520.

Couly, G. F., and Le Douarin, N. M. (1985). Mapping of the early neural primordium in quail-chick chimeras. I. Developmental relationships between placodes, facial ectoderm, prosen-cephalon. *Dev Biol* **110**, pp. 422–439.

Cvekl, A., Yang, Y., Chauhan, B. K., and Cveklova, K. (2004). Regulation of gene expression by Pax6 in ocular cells: a case of tissue-preferred expression of crystallins in lens. *Int J Dev Biol* **48**, pp. 829–844.

Dahl, E., Koseki, H., and Balling, R. (1997). Pax genes and organogenesis. *BioEssays* **19**, pp. 755–765.

Damas, H. (1951). Observations sur le developpement des ganglions craniens chez *Lampetra flu-viatilis* (L.). *Arch Biol* **62**, pp. 65–95.

Dambly-Chaudière, C., Sapède, D., Soubiran, F., Decorde, K., Gompel, N., and Ghysen, A. (2004). The lateral line of zebrafish: a model system for the analysis of morphogenesis and neural development in vertebrates. *Biol Cell* **95**, pp. 579–584.

Dambly-Chaudiere, C., Cubedo, N., and Ghysen, A. (2007). Control of cell migration in the de-velopment of the posterior lateral line: antagonistic interactions between the chemokine receptors CXCR4 and CXCR7/RDC1. *BMC Dev Biol* **7**, p. 23.

D'Amico-Martel, A., and Noden, D. M. (1983). Contributions of placodal and neural crest cells to avian cranial peripheral ganglia. *Am J Anat* **166**, pp. 445–468.

Davis, R. J., Shen, W., Sandler, Y. I., Amoui, M., Purcell, P., Maas, R., Ou, C. N., Vogel, H., Beaudet, A. L., and Mardon, G. (2001). Dach1 mutant mice bear no gross abnormalities in eye, limb, and brain development and exhibit postnatal lethality. *Mol Cell Biol* **21**, pp. 1484–1490.

Davis, R. J., Shen, W., Heanue, T. A., and Mardon, G. (1999). Mouse *Dach*, a homologue of *Dro-sophila dachshund*, is expressed in the developing retina, brain and limbs. *Dev Genes Evol* **209**, pp. 526–536.

Deitcher, D. L., Fekete, D. M., and Cepko, C. L. (1994). Asymmetric expression of a novel homeo-box gene in vertebrate sensory organs. *J Neurosci* **14**, pp. 486–498.

Depew, M. J., Liu, J. K., Long, J. E., Presley, R., Meneses, J. J., Pedersen, R. A., and Rubenstein, J. L. (1999). *Dlx5* regulates regional development of the branchial arches and sensory cap-sules. *Development* **126**, pp. 3831–3846.

Devine, C. A., Sbrogna, J. L., Guner, B., Osgood, M., Shen, M. C., and Karlstrom, R. O. (2009). A dynamic Gli code interprets Hh signals to regulate induction, patterning, and endocrine cell specification in the zebrafish pituitary. *Dev Biol* **326**, pp. 143–154.

Dijkgraff, S. (1963). The functioning and significance of the lateral-line organs. *Biol Rev* **38**, pp. 51–105.

Donner, A. L., and Maas, R. L. (2004). Conservation and non-conservationof genetic pathways in eye specification. *Int J Dev Biol* **48**, pp. 743–753.

Donner, A. L., Episkopou, V., and Maas, R. L. (2007). Sox2 and Pou2f1 interact to control lens and olfactory placode development. *Dev Biol* **303**, pp. 784–799.

Drysdale, T. A., and Elinson, R. P. (1991). Development of the Xenopus laevis hatching gland and its relationship to surface ectoderm patterning. *Development* **111**, pp. 469–478.

Dubois, L., Bally-Cuif, L., Crozatier, M., Moreau, J., Paquereau, L., and Vincent, A. (1998). XCoe2, a transcription factor of the Col/Olf-1/EBF family involved in the specification of primary neurons in *Xenopus. Curr Biol* **8**, pp. 199–209.

Dude, C. M., Kuan, C. Y., Bradshaw, J. R., Greene, N. D., Relaix, F., Stark, M. R., and Baker, C. V. (2009). Activation of Pax3 target genes is necessary but not sufficient for neurogenesis in the ophthalmic trigeminal placode. *Dev Biol* **326**, pp. 314–326.

Duggan, C. D., DeMaria, S., Baudhuin, A., Stafford, D., and Ngai, J. (2008). Foxg1 is required for development of the vertebrate olfactory system. *J Neurosci* **28**, pp. 5229–5239.

Dutta, S., Dietrich, J. E., Aspock, G., Burdine, R. D., Schier, A., Westerfield, M., and Varga, Z. M. (2005). Pitx3 defines an equivalence domain for lens and anterior pituitary placode. *Development* **132**, pp. 1579–1590.

Eagleson, G. W., Jenks, B. G., and Van Overbeeke, A. P. (1986). The pituitary adrenocorticotropes originate from neural ridge tissue in Xenopus laevis. *J Embryol Exp Morphol* **95**, pp. 1–14.

Eagleson, G., Ferreiro, B., and Harris, W. A. (1995). Fate of the anterior neural ridge and the morphogenesis of the Xenopus forebrain. *J Neurobiol* **28**, pp. 146–158.

Ede, D. A., and Kelly, W. A. (1964). Developmental abnormalities in the head region of the *talpid*3 mutant of the fowl. *J Embryol Exp Morph* **12**, pp. 161–182.

elAmraoui, A., and Dubois, P. M. (1993). Experimental evidence for the early commitment of the presumptive adenohypophysis. *Neuroendocrinology* **58**, pp. 609–615.

Ellies, D. L., Stock, D. W., Hatch, G., Giroux, G., Weiss, K. M., and Ekker, M. (1997). Relationship between the genomic organization and the overlapping embryonic expression patterns of the zebrafish *dlx* genes. *Genomics* **45**, pp. 580–590.

Epstein, D. J., Vekemans, M., and Gros, P. (1991). Splotch (Sp2H), a mutation affecting development of the mouse neural tube, shows a deletion within the paired homeodomain of Pax-3. *Cell* **67**, pp. 767–774.

Ericson, J., Norlin, S., Jessell, T. M., and Edlund, T. (1998). Integrated FGF and BMP signalling controls the progression of progenitor cell differentiation and the emergence of pattern in the embryonic anterior pituitary. *Development* **125**, pp. 1005–1015.

Esteve, P., and Bovolenta, P. (1999). *cSix4*, a member of the six gene family of transcription factors, is expressed during placode and somite development. *Mech Dev* **85**, pp. 161–165.

Farbman, A. I. (1994). Developmental biology of olfactory sensory neurons. *Semin Cell Biol* **5**, pp. 3–10.

Fekete, D. M., and Wu, D. K. (2002). Revisiting cell fate specification in the inner ear. *Curr Opin Neurobiol* **12**, pp. 35–42.

Fode, C., Gradwohl, G., Morin, X., Dierich, A., LeMeur, M., Goridis, C., and Guillemot, F. (1998). The bHLH protein NEUROGENIN 2 is a determination factor for epibranchial placode-derived sensory neurons. *Neuron* **20**, pp. 483–494.

Fritzsch, B., Beisel, K. W., Jones, K., Farinas, I., Maklad, A., Lee, J., and Reichardt, L. F. (2002). Development and evolution of inner ear sensory epithelia and their innervation. *J Neurobiol* **53**, pp. 143–156.

Fritzsch, B., Pauley, S., and Beisel, K. W. (2006). Cells, molecules and morphogenesis: the making of the vertebrate ear. *Brain Res* **1091**, pp. 151–171.

Furuta, Y., and Hogan, B. L. M. (1998). BMP4 is essential for lens induction in the mouse embryo. *Genes Dev* **12**, pp. 3764–3775.

Gage, P. J., and Camper, S. A. (1997). Pituitary homeobox 2, a novel member of the *bicoid*-related family of homeobox genes, is a potential regulator of anterior structure formation. *Hum Mol Genet* **6**, pp. 457–464.

Gehring, W. J., and Ikeo, K. (1999). Pax 6: mastering eye morphogenesis and eye evolution. *Trends Genet* **15**, pp. 371–377.

George, K. M., Leonard, M. W., Roth, M. E., Lieuw, K. H., Kioussis, D., Grosveld, F., and Engel, J. D. (1994). Embryonic expression and cloning of the murine GATA-3 gene. *Development* **120**, pp. 2673–2686.

Gehring, W. J., and Ikeo, K. (1999). Pax 6-mastering eye morphogenesis and eye evolution. *Trends Genet* **15**, pp. 371–377.

Ghanbari, H., Seo, H. C., Fjose, A., and Brändli, A. W. (2001). Molecular cloning and embryonic expression of Xenopus Six homeobox genes. *Mech Dev* **101**, pp. 271–277.

Ghysen, A., Dambly-Chaudiere, C. (2004). Development of the zebrafish lateral line. *Curr Opin Neurobiol* **14**, pp. 67–73.

Ghysen, A., and Dambly-Chaudiere, C. (2007). The lateral line microcosmos. *Genes Dev* **21**, pp. 2118–2130.

Gibbs, A. M. (2004). Lateral line receptors: where do they come from developmentally and where is our research going? *Brain Behav Evol* **64**, pp. 163–181.

Giraldez, F. (1998). Regionalized organizing activity of the neural tube revealed by the regulation of *lmx1* in the otic vesicle. *Dev Biol* **203**, pp. 189–200.

Glasgow, E., and Tomarev, S. I. (1998). Restricted expression of the homeobox gene *prox 1* in developing zebrafish. *Mech Dev* **76**, pp. 175–178.

Glasgow, E., Karavanov, A. A., and Dawid, I. B. (1997). Neuronal and neuroendocrine expression of *lim3*, a LIM class homeobox gene, is altered in mutant zebrafish with axial signalling defects. *Dev Biol* **192**, pp. 405–419.

Glavic, A., Maris Honoré, S., Gloria Feijóo, C., Bastidas, F., Allende, M. L., Mayor, R. (2004). Role of BMP signaling and the homeoprotein Iroquois in the specification of the cranial placodal field. *Dev Biol* **272**, pp. 89–103.

Grainger, R. M. (1996). New perspectives on embryonic lens induction. *Semin Cell Dev Biol* **7**, pp. 149–155.

Grifone, R., Demignon, J., Houbron, C., Souil, E., Niro, C., Seller, M. J., Hamard, G., and Maire, P. (2005). Six1 and Six4 homeoproteins are required for Pax3 and Mrf expression during myogenesis in the mouse embryo. *Development* **132**, pp. 2235–2249.

Grindley, J. C., Davidson, D. R., and Hill, R. E. (1995). The role of *Pax-6* in eye and nasal development. *Development* **121**, pp. 1433–1442.

Groves, A. K., and Bronner-Fraser, M. (2000). Competence, specification and commitment in otic placode induction. *Development* **127**, pp. 3489–3499.

Guo, S., Brush, J., Teraoka, H., Goddard, A., Wilson, S. W., Mullins, M. C., and Rosenthal, A. (1999). Development of noradrenergic neurons in the zebrafish hindbrain requires BMP, FGF8, and the homeodomain protein soulless/Phox2a. *Neuron* **24**, pp. 555–566.

Hamburger, V. (1961). Experimental analysis of the dual origin of the trigeminal ganglion of the chick embryo. *J Exp Zool* **148**, pp. 91–117.

Hamburger, V., and Hamilton, H. L. (1992). A series of normal stages in the development of the chick embryo. *Dev Dyn* **195**, pp. 231–272.

Hammond, K. L., Hanson, I. M., Brown, A. G., Lettice, L. A., and Hill, R. E. (1998). Mammalian and *Drosophila dachshund* genes are related to the *Ski* proto-oncogene and are expressed in eye and limb. *Mech Dev* **74**, pp. 121–131.

Hans, S., Christison, J., Liu, D., and Westerfield, M. (2007). Fgf-dependent otic induction requires competence provided by Foxi1 and Dlx3b. *BMC Dev Biol* **7**, p. 5.

Hans, S., Liu, D., and Westerfield, M. (2004). Pax8 and Pax2a function synergistically in otic specification, downstream of the Foxi1 and Dlx3b transcription factors. *Development* **131**, pp. 5091–5102.

Hatini, V., Ye, X., Balas, G., and Lai, E. (1999). Dynamics of placodal lineage development revealed by targeted transgene expression. *Dev Dyn* **215**, pp. 332–343.

Hausen, P., and Riebesell, M. (1991). The early development of *Xenopus laevis*. Verlag Der Zeitschrift fur Naturforschung. Tubingen – Germany.

Heller, N., and Brändli, A. W. (1997). *Xenopus Pax-2* displays multiple splice forms during embryogenesis and pronephric kidney development. *Mech Dev* **69**, pp. 83–104.

Heller, N., and Brändli, A. W. (1999). Xenopus Pax-2/5/8 orthologues: novel insights into Pax gene evolution and identification of Pax-8 as the earliest marker for otic and pronephric cell lineages. *Dev Genet* **24**, pp. 208–219.

Herbrand, H., Guthrie, S., Hadrys, T., Hoffmann, S., Arnold, H. H., Rinkwitz-Brandt, S., and Bober, E. (1998). Two regulatory genes, *cNkx5-1* and *cPax2*, show different responses to local signals during otic placode and vesicle formation in the chick embryo. *Development* **125**, pp. 645–654.

Hermesz, E., Mackem, S., and Mahon, K. A. (1996). *Rpx*: a novel anterior-restricted homeobox gene progressively activated in the prechordal plate, anterior neural plate and Rathke's pouch of the mouse embryo. *Development* **122**, pp. 41–52.

Herzog, W., Zeng, X., Lele, Z., Sonntag, C., Ting, J. W., Chang, C. Y., and Hammerschmidt, M. (2003). Adenohypophysis formation in the zebrafish and its dependence on sonic hedgehog. *Dev Biol* **25**, 36–49.

Hidalgo-Sánchez, M., Alvarado-Mallart, R., and Alvarez, I. S. (2000). *Pax2, Otx2, Gbx2* and *Fgf8* expression in early otic vesicle development. *Mech Dev* **95**, pp. 225–229.

Higgs, D. M., and Burd, G. D. (1999). The role of brain in metamorphosis of the olfactory epithelium in the frog, Xenopus laevis. *Dev Brian Res* **118**, pp. 185–195.

Hirsch, N., and Harris, W. A. (1997). *Xenopus* Pax-6 and retinal development. *J Neurobiol* **32**, pp. 45–61.

Hogan, B. L., Horsburgh, G., Cohen, J., Hetherington, C. M., Fisher, G., and Lyon, M. F. (1986). Small eyes (Sey): a homozygous lethal mutation on chromosome 2 which affects the differentiation of both lens and nasal placodes in the mouse. *J Embryol Exp Morphol* **97**, pp. 95–110.

Hollemann, T., and Pieler, T. (1999). *Xpitx-1*: a homeobox gene expressed during pituitary and cement gland formation of *Xenopus* embryos. *Mech Dev* **88**, pp. 249–252.

Honoré, S. M., Aybar, M. J., and Mayor, R. (2003). Sox10 is required for the early development of the prospective neural crest in Xenopus embryos. *Dev Biol* **260**, pp. 79–96.

Huang, X., Hong, C. S., O'Donnell, M., and Saint-Jeannet, J.-P. (2005). The doublesex-related gene, XDmrt4, is required for neurogenesis in the olfactory system. *Proc Natl Acad Sci USA* **102**, pp. 11349–11354.

Huang, X., and Saint-Jeannet, J.-P. (2004). Induction of the neural crest and the opportunities of life on the edge. *Dev Biol* **275**, pp. 1–11.

Hong, C.-S., and Saint-Jeannet, J.-P. (2007). The activity of Pax3 and Zic1 regulates three distinct cell fates at the neural plate border. *Mol Biol Cell* **18**, pp. 2192–2202.

Hong, C.-S., Park, B.-Y., and Saint-Jeannet, J.-P. (2008). Fgf8a induces neural crest indirectly through the activation of Wnt8 in the paraxial mesoderm. *Development* **135**, pp. 3903–3910.

Ishibashi, S., and Yasuda, K. (2001). Distinct roles of maf genes during Xenopus lens development. *Mech Dev* **101**, pp. 155–166.

Itoh, M., Kudoh, T., Dedekian, M., Kim, C. H., and Chitnis, A. B. (2002). A role for iro1 and iro7 in the establishment of an anteroposterior compartment of the ectoderm adjacent to the midbrain-hindbrain boundary. *Development* **129**, pp. 2317–2327

Jacobson, A. G. (1963a). The determination and positioning of the nose, lens and ear. I. Interactions within the ectoderm, and between the ectoderm and underlying tissues. *J Exp Zool* **154**, pp. 273–283.

Jacobson, A. G. (1963b). Determination and positioning of the nose, lens and ear. II. The role of the endoderm. *J Exp Zool* **154**, pp. 285–291.

Jacobson, A. G. (1963c). The determination and positioning of the nose, lens and ear. III. Effects of reversing the antero-posterior axis of epidermis, neural plate and neural fold. *J Exp Zool* **154**, pp. 293–303.

Jacobson, A. G. (1966). Inductive processes in embryonic development. *Science* **152**, pp. 25–34.

Jean, D., Bernier, G., and Gruss, P. (1999). *Six6* (*Optx2*) is a novel murine *Six3*-related homeobox gene that demarcates the presumptive pituitary/hypothalamic axis and the ventral optic stalk. *Mech Dev* **84**, pp. 31–40.

Kablar, B., Vignali, R., Menotti, L., Pannese, M., Andreazzoli, M., Polo, C., Giribaldi, M. G., Boncinelli, E., and Barsacchi, G. (1996). *Xotx* genes in the developing brain of *Xenopus laevis*. *Mech Dev* **55**, pp. 145–158.

Kamachi, Y., Uchikawa, M., Collignon, J., Lovell-Badge, R., and Kondoh, H. (1998). Involvement of Sox1, 2 and 3 in the early and subsequent molecular events of lens induction. *Development* **125**, pp. 2521–2532.

Karlstrom, R. O., Talbot, W. S., and Schier, A. F. (1999). Comparative synteny cloning of zebrafish *you-too*: mutations in the Hedgehog target *gli2* affect ventral forebrain patterning. *Genes Dev* **13**, pp. 388–393.

Kataoka, H., Ochi, M., Enomoto, K., and Yamaguchi, A. (2000). Cloning and embryonic expression patterns of the zebrafish Runt domain genes, *runxa* and *runxb*. *Mech Dev* **98**, pp. 139–143.

Kawahara, A., and Dawid, I. B. (2002). Developmental expression of zebrafish emx1 during early embryogenesis. *Gene Expr Patterns* **2**, pp. 201–206.

Kawakami, K., Sato, S., Ozaki, H., and Ikeda, K. (2000). Six family genes-structure and function as transcription factors and their roles in development. *Bioessays* **22**, pp. 616–626.

Kawamura, K., Kouki, T., Kawahara, G., and Kikuyama, S. (2002). Hypophyseal development in vertebrates from amphibians to mammals. *Gen Comp Endocrinol* **126**, pp. 130–135.

Kawauchi, S., Takahashi, S., Nakajima, O., Ogino, H., Morita, M., Nishizawa, M., Yasuda, K., and Yamamoto, M. (1999). Regulation of lens fiber cell differentiation by transcription factor c-Maf. *J Biol Chem* **274**, pp. 19254–19260.

Kazanskaya, O. V., Severtzova, E. A., Barth, K. A., Ermakova, G. V., Lukyanov, S. A., Benyumov, A. O., Pannese, M., Boncinelli, E., Wilson, S. W., and Zaraisky, A. G. (1997). Anf: a novel class of vertebrate homeobox genes expressed at the anterior end of the main embryonic axis. *Gene* **200**, pp. 25–34.

Kenyon, K. L., Moody, S. A., and Jamrich, M. (1999). A novel *fork head* gene mediates early steps during *Xenopus* lens formation. *Development* **126**, pp. 5107–5116.

Kil, S. H., Streit, A., Brown, S. T., Agrawal, N., Collazo, A., Zile, M. H., and Groves, A. K. (2005). Distinct roles for hindbrain and paraxial mesoderm in the induction and patterning of the inner ear revealed by a study of vitamin-A-deficient quail. *Dev Biol* **285**, pp. 252–271.

Kim, J. I., Li, T., Ho, I. C., Grusby, M. J., and Glimcher, L. H. (1999). Requirement for the c-Maf transcription factor in crystallin gene regulation and lens development. *Proc Natl Acad Sci USA* **96**, pp. 3781–3785.

Kimmel, C. B., Ballard, W. W., Kimmel, S. R., Ullman, B., and Schilling, T. F. (1995). Stages of embryonic development of the zebrafish. *Dev Dyn* **203**, pp. 253–310.

Klein, S. L., and Graziadei, P. P. (1983). The differentiation of the olfactory placode in Xenopus laevis: a light and electron microscope study. *J Comp Neurol* **217**, pp. 17–30.

Knaut, H., Blader, P., Strahle, U., and Schier, A. F. (2005). Assembly of trigeminal sensory ganglia by chemokine signaling. *Neuron* **47**, pp. 653–666.

Knecht, A. K., and Bronner-Fraser, M. (2002). Induction of the neural crest: a multigene process. *Nat Rev Genet* **3**, pp. 453–461.

Kobayashi, M., Osanai, H., Kawakami, K., and Yamamoto, M. (2000). Expression of three zebrafish *Six4* genes in the cranial sensory placodes and the developing somites. *Mech Dev* **98**, pp. 151–155.

Kobayashi, M., Toyama, R., Takeda, H., Dawid, I. B., and Kawakami, K. (1998). Overexpression of the forebrain-specific homeobox gene *six3* induces rostral forebrain enlargement in zebrafish. *Development* **125**, pp. 2973–2982.

Konishi, Y., Ikeda, K., Iwakura, Y., and Kawakami, K. (2006). Six1 and Six4 promote survival of sensory neurons during early trigeminal gangliogenesis. *Brain Res* **1116**, pp. 93–102.

Korzh, V., Edlund, T., and Thor, S. (1993). Zebrafish primary neurons initiate expression of the LIM homeodomain protein Isl-1 at the end of gastrulation. *Development* **118**, pp. 417–425.

Kozlowski, D. J., Murakami, T., Ho, R. K., and Weinberg, E. S. (1997). Regional cell movement and tissue patterning in the zebrafish embryo revealed by fate mapping with caged fluorescein. *Biochem Cell Biol* **75**, pp. 551–562.

Kozlowski, D. J., Whitfield, T. T., Hukriede, N. A., Lam, W. K., Weinberg, E. S. (2005). The zebrafish dog-eared mutation disrupts eya1, a gene required for cell survival and differentiation in the inner ear and lateral line. *Dev Biol* **277**, pp. 27–41.

Krauss, S., Johansen, T., Korzh, V., and Fjose, A. (1991). Expression of the zebrafish paired box gene *pax[zf-b]* during early neurogenesis. *Development* **113**, pp. 1193–1206.

Kumar, J.-P. (2009). The sine oculis homeobox (SIX) family of transcription factors as regulators of development and disease. *Cell Mol Life Sci* **66**, pp. 565–583.

Kuratani, S. C., and Wall, N. A. (1992). Expression of Hox 2.1 protein in restricted populations of neural crest cells and pharyngeal ectoderm. *Dev Dyn* **195**, pp. 15–28.

LaBonne, C., and Bronner-Fraser, M. (1998). Neural crest induction in Xenopus: evidence for a two-signal model. *Development* **125**, pp. 2403–2414.

Laclef, C., Souil, E., Demignon, J., and Maire, P. (2003). Thymus, kidney and craniofacial abnormalities in Six 1 deficient mice. *Mech Dev* **120**, pp. 669–679.

Ladher, R. K., Anakwe, K. U., Gurney, A. L., Schoenwolf, G. C., and Francis-West, PH. (2000). Identification of synergistic signals initiating inner ear development. *Science* **290**, pp. 1965–1967.

Ladher, R. K., Wright, T. J., Moon, A. M., Mansour, S. L., and Schoenwolf, G. C. (2005). FGF8 initiates inner ear induction in chick and mouse. *Genes & Dev* **19**, pp. 603–613.

Ladher, R. K., O'Neill, P., and Begbie, J. (2010). From shared lineage to distinct functions: development of the inner ear and epibranchial placodes. *Development* **137**, pp. 1777–1785.

Lanctôt, C., Lamolet, B., and Drouin, J. (1997). The bicoid-related homeoprotein Ptx1 defines the most anterior domain of the embryo and differentiates posterior from anterior lateral mesoderm. *Development* **124**, pp. 2807–2817.

Lang, R. A. (2004). Pathways regulating lens induction in the mouse. *Int J Dev Biol* **48**, pp. 783–791.

Lassiter, R. N., Dude, C. M., Reynolds, S. B., Winters, N. I., Baker, C. V., and Stark, M. R. (2007). Canonical Wnt signaling is required for ophthalmic trigeminal placode cell fate determination and maintenance. *Dev Biol* **308**, 392–406.

Lecaudey, V., Anselme, I., Dildrop, R., Rüther, U., and Schneider-Maunoury, S. (2005). Expression of the zebrafish *Iroquois* genes during early nervous system formation and patterning. *J Comp Neurol* **492**, pp. 289–302.

Ledent, V. (2002). Postembryonic development of the posterior lateral line in zebrafish. *Development* **129**, pp. 597–604.

Le Douarin, N. M., Fontaine-Perus, J., and Couly, G (1986). Cephalic ectodermal placodes and neurogenesis. *Trends Neurosci* **9**, pp. 175–180.

Lee, S. A., Shen, E. L., Fiser, A., Sali, A., and Guo, S. (2003). The zebrafish forkhead transcription factor Foxi1 specifies epibranchial placode-derived sensory neurons. *Development* **130**, pp. 2669–2679.

Leger, S., and Brand, M. (2002). Fgf8 and Fgf3 are required for zebrafish ear placode induction, maintenance and inner ear patterning. *Mech Dev* **119**, pp. 91–108.

Leimeister, C., Externbrink, A., Klamt, B., and Gessler, M. (1999). *Hey* genes: a novel subfamily of *hairy*- and *Enhancer of split*related genes specifically expressed during mouse embryogenesis. *Mech Dev* **85**, pp. 173–177.

León, Y., Miner, C., Represa, J., and Giraldez, F. (1992). Myb p75 oncoprotein is expressed in developing otic and epibranchial placodes. *Dev Biol* **153**, pp. 407–410.

Lewis, K. E., Drossopoulou, G., Paton, I. R., Morrice, D. R., Robertson, K. E., Burt, D. W., Ingham, P. W., and Tickle, C. (1999). Expression of ptc and gli genes in talpid3 suggests bifurcation in Shh pathway. *Development* **126**, pp. 2397–2407.

Li, H. S., Yang, J. M., Jacobson, R. D., Pasko, D., and Sundin, O. (1994). *Pax-6* is first expressed in a region of ectoderm anterior to the early neural plate: implications for stepwise determination of the lens. *Dev Biol* **162**, pp. 181–194.

Li, X., Cvekl, A., Bassnett, S., and Piatigorsky, J. (1997). Lenspreferred activity of chicken d1- and d2-crystallin enhancers in transgenic mice and evidence for retinoic acid-responsive regulation of the d1-crystallin gene. *Dev Genet* **20**, pp. 258–266.

Li, X., Oghi, K. A., Zhang, J., Krones, A., Bush, K. T., Glass, C. K., Nigam, S. K., Aggarwal, A. K., Maas, R., Rose, D. W., and Rosenfeld, M. G. (2003). Eya protein phosphatase activity regulates Six1-Dach-Eya transcriptional effects in mammalian organogenesis. *Nature* **426**, pp. 247–254.

Li, H., Liu, H., Sage, C., Huang, M., Chen, Z. Y., and Heller, S. (2004). Islet-1 expression in the developing chicken inner ear. *J Comp Neurol* **477**, pp. 1–10.

Li, Q., Zhou, X., Guo, Y., Shang, X., Chen, H., Lu, H., Cheng, H., and Zhou, R. (2008). Nuclear localization, DNA binding and restricted expression in neural and germ cells of zebrafish Dmrt3. *Biol Cell* **100**, pp. 453–463.

Lilleväli, K., Haugas, M., Pituello, F., and Salminen, M. (2007). Comparative analysis of Gata3 and Gata2 expression during chicken inner ear development. *Dev Dyn* **236**, pp. 306–313.

Litsiou, A., Hanson, S., and Streit, A. (2005). A balance of FGF, BMP and WNT signalling positions the future placode territory in the head. *Development* **132**, pp. 4051–4062.

Liu, D., Chu, H., Maves, L., Yan, Y. L., Morcos, P. A., Postlethwait, J. H., and Westerfield, M. (2003). Fgf3 and Fgf8 dependent and independent transcription factors are required for otic placode specification. *Development* **130**, pp. 2213–2224.

Logan, C., Wingate, R. J., McKay, I. J., and Lumsden, A. (1998). *Tlx-1* and *Tlx-3* homeobox gene expression in cranial sensory ganglia and hindbrain of the chick embryo: markers of patterned connectivity. *J Neurosci* **18**, pp. 5389–5402.

Lovicu, F. J., and McAvoy, J. W. (2005). Growth factor regulation of lens development. *Dev Biol* **280**, pp. 1–14.

Ma, E. Y., and Raible, D. W. (2009). Signaling pathways regulating zebrafish lateral line development. *Curr Biol* **19**, pp. R381–R386.

Ma, Q., Chen, Z., del Barco Barrantes, I., de la Pompa, J. L., and Anderson, D. J. (1998). *neurogenin1* is essential for the determination of neuronal precursors for proximal cranial sensory ganglia. *Neuron* **20**, pp. 469–482.

MacKenzie, A., Ferguson, M. W., and Sharpe, P. T. (1991). *Hox-7* expression during murine craniofacial development. *Development* **113**, pp. 601–611.

Mackereth, M. D., Kwak, S. J., Fritz, A., and Riley, B. B. (2005). Zebrafish Pax8 is required for otic placode induction and plays a redundant role with Pax2 genes in the maintenance of the otic placode. *Development* **132**, pp. 371–382.

Maden, M., Blentic, A., Reijntjes, S., Sequin, S., Gale, E., and Graham, A. (2007). Retinoic acid is required fro specification of the ventral eye field and for Rathke's pouch in the avian embryo. *Int J Dev Biol* **51**, pp. 191–200.

Maier, E., and Gunhaga, L. (2009). Dynamic expression of neurogenic markers in the developing chick olfactory epithelium. *Dev Dyn* **238**, pp. 1617–1625.

Mallamaci, A., Di Blas, E., Briata, P., Boncinelli, E., and Corte, G. (1996). OTX2 homeoprotein in the developing central nervous system and migratory cells of the olfactory area. *Mech Dev* **58**, pp. 165–178.

Mallamaci, A., Iannone, R., Briata, P., Pintonello, L., Mercurio, S., Boncinelli, E., and Corte, G. (1998). EMX2 protein in the developing mouse brain and olfactory area. *Mech Dev* **77**, pp. 165–172.

Mansouri, A., Chowdhury, K., and Gruss, P. (1998). Follicular cells of the thyroid gland require Pax8 gene function. *Nat Genet* **19**, pp. 87–90.

Marchant, L., Linker, C., Ruiz, P., Guerrero, N., and Mayor, R. (1998). The inductive properties of mesoderm suggest that the neural crest cells are specified by a BMP gradient. *Dev Biol* **198**, pp. 319–329.

Maroon, H., Walshe, J., Mahmood, R., Kiefer, P., Dickson, C., and Mason, I. (2002). Fgf3 and Fgf8 are required together for formation of the otic placode and vesicle. *Development* **129**, pp. 2099–2108.

Mathers, P. H., Miller, A., Doniach, T., Dirksen, M. L., and Jamrich, M. (1995). Initiation of anterior head-specific gene expression in uncommitted ectoderm of *Xenopus laevis* by ammonium chloride. *Dev Biol* **171**, pp. 641–654.

Mayor, R., Guerrero, N., and Martinez, C. (1997). Role of FGF and noggin in neural crest induction. *Dev Biol* **189**, pp. 1–12.

McCabe, K. L., and Bronner-Fraser, M. (2008). Essential role for PDGF signaling in ophthalmic trigeminal placode induction. *Development* **135**, pp. 1863–1874.

McCabe, K. L., and Bronner-Fraser, M. (2009). Molecular and tissue interactions governing induction of cranial ectodermal placodes. *Dev Biol* **332**, pp. 189–195.

Mendonsa, E. S., and Riley, B. B. (1999). Genetic analysis of tissue interactions required for otic placode induction in the zebrafish. *Dev Biol* **206**, pp. 100–112.

Meulemans, D., and Bronner-Fraser, M. (2004). Gene-regulatory interactions in neural crest evolution and development. *Dev Cell* **7**, pp. 291–299.

Millimaki, B. B., Sweet, E. M., Dhason, M. S., and Riley, B. B. (2007). Zebrafish atoh1 genes: classic proneural activity in the inner ear and regulation by Fgf and Notch. *Development* **134**, pp. 295–305.

Mishima, N., and Tomarev, S. (1998). Chicken *Eyes absent 2* gene: isolation and expression pattern during development. *Int J Dev Biol* **42**, pp. 1109–1115.

Miyasaka, N., Knaut, H., and Yoshihara, Y. (2007). Cxcl12/Cxcr4 chemokine signaling is required for placode assembly and sensory neurons pathfinding in the zebrafish olfactory system. *Development* **134**, pp. 2459–2468.

Mizuseki, K., Kishi, M., Matsui, M., Nakanishi, S., and Sasai, Y. (1998). *Xenopus* Zic-related-1 and Sox-2, two factors induced by chordin, have distinct activities in the initiation of neural induction. *Development* **125**, pp. 579–587.

Molenaar, M., Brian, E., Roose, J., Clevers, H., and Destre´e, O. (2000). Differential expression of the Groucho-related genes 4 and 5 during early development of *Xenopus laevis. Mech Dev* **91**, pp. 311–315.

Monsoro-Burq, A. H., Fletcher, R. B., and Harland, R. M. (2003). Neural crest induction by paraxial mesoderm in Xenopus embryos requires FGF signals. *Development* **130**, pp. 3111–3124.

Monsoro-Burq, A. H., Wang, E., and Harland, R. (2005). Msx1 and Pax3 cooperate to mediate FGF8 and WNT signals during Xenopus neural crest induction. *Dev Cell* **8**, pp. 167–178.

Monaghan, A. P., Davidson, D. R., Sime, C., Graham, E., Baldock, R., Bhattacharya, S. S., and Hill, R. E. (1991). The *Msh*-like homeobox genes define domains in the developing vertebrate eye. *Development* **112**, pp. 1053–1061.

Monaghan, A. P., Grau, E., Bock, D., and Schutz, G. (1995). The mouse homolog of the orphan nuclear receptor *tailless* is expressed in the developing forebrain. *Development* **121**, pp. 839–853.

Montgomery, J., Carton, G., Voigt, R., Baker, C., and Diebel, C. (2000). Sensory processing of water currents by fishes. *Philos Trans R Soc Lond B Biol Sci* **355**, pp. 1325–1327.

Mori, M., Ghyselinck, N. B., Chambon, P., and Mark, M. (2001). Systematic immunolocalization of retinoid receptors in developing and adult mouse eyes. *Invest Ophthalmol Vis Sci* **42**, pp. 1312–1318.

Morin, X., Cremer, H., Hirsch, M.-R., Kapur, R. P., Goridis, C., and Brunet, J.-F. (1997). Defects in sensory and autonomic ganglia and absence of locus coeruleus in mice deficient for the homeobox gene *Phox2a Neuron* **18**, pp. 411–423.

Morsli, H., Tuorto, F., Choo, D., Postiglione, M. P., Simeone, A., and Wu, D. K. (1999). *Otx1* and *Otx2* activities are required for the normal development of the mouse inner ear. *Development* **126**, pp. 2335–2343.

Münchberg, S. R., Ober, E. A., and Steinbesser, H. (1999). Expression of the Ets transcription factors erm and pea3 in early zebrafish development. *Mech Dev* **88**, pp. 233–236.

Mucchielli, M. L., Martinez, S., Pattyn, A., Goridis, C., and Brunet, J. F. (1996). *Otlx2*, an *Otx*-related homeobox gene expressed in the pituitary gland and in a restricted pattern in the forebrain. *Mol Cell Neurosci* **8**, pp. 258–271.

Narayanan, C. H., and Narayanan, Y. (1978). Determination of the embryonic origin of the mesencephalic nucleus of the trigeminal nerve in birds. *J Embryol Exp Morphol* **43**, pp. 85–105.

Narayanan, C. H., and Narayanan, Y. (1980). Neural crest and placodal contributions in the development of the glossopharyngeal-vagal complex in the chick. *Anat Rec* **196**, pp. 71–82.

Neave, B., Holder, N., and Patient, R. (1997). A graded response to Bmp-4 spatially coordinates patterning of the mesoderm and ectoderm in the zebrafish. *Mech Dev* **62**, pp. 183–185.

Nechiporuk, A., and Raible, D. W. (2008). FGF-dependent mechanosensory organ patterning in zebrafish. *Science* **320**, pp. 1774–1777.

Nechiporuk, A., Linbo, T., Poss, K. D., and Raible, D. W. (2007). Specification of epibranchial placodes in zebrafish. *Development* **134**, pp. 611–623.

Nguyen, V. H., Schmid, B., Trout, J., Connors, S. A., Ekker, M., and Mullins, M. C. (1998). Ventral and lateral regions of the zebrafish gastrula, including the neural crest progenitors, are established by a bmp2b/swirl pathway of genes. *Dev Biol* **199**, pp. 93–110.

Nikaido, M., Doi, K., Shimizu, T., Hibi, M., Kikuchi, Y., and Yamasu, K. (2007). Initial specification of the epibranchial placode in zebrafish embryos depends on the fibroblast growth factor signal. *Dev Dyn* **236**, pp. 564–571.

Nichols, D. H., Pauley, S., Jahan, I., Beisel, K. W., Millen, K. J., and Fritzsch, B. (2008). Lmx1a is required for segregation of sensory epithelia and normal ear histogenesis and morphogenesis. *Cell Tissue Res* **334**, pp. 339–358.

Nichols, D. H. (1986). Mesenchyme formation from the trigeminal placodes of the mouse embryo. *Am J Anat* **176**, pp. 19–31.

Nieuwkoop, P. D., and Faber, J. (1967). Normal table of *Xenopus laevis* (Daudin), Amsterdam, The Netherlands: North Holland Publishing Company.

Nikaido, M., Doi, K., Shimizu, T., Hibi, M., Kikuchi, Y., and Yamasu, K. (2007). Initial specification of the epibranchial placode in zebrafish embryos depends on the fibroblast growth factor signal. *Dev Dyn* **236**, pp. 564–571.

Nishiguchi, S., Wood, H., Kondoh, H., Lovell-Badge, R., and Episkopou, V. (1998). *Sox1* directly regulates the g-crystallin genes and is essential for lens development in mice. *Genes Dev* **12**, pp. 776–781.

Nissen, R. M., Yan, J., Amsterdam, A., Hopkins, N., and Burgess, S. M. (2003). Zebrafish foxi one modulates cellular responses to Fgf signaling required for the integrity of ear and jaw patterning. *Development* **130**, pp. 2543–2554.

Noramly, S., and Grainger, R. M. (2002). Determination of the embryonic inner ear. *J. Neurobiol* **53**, pp. 100–128.

Nornes, H. O., Dressler, G. R., Knapik, E. W., Deutsch, U., and Gruss, P. (1990). Spatially and temporally restricted expression of *Pax2* during murine neurogenesis. *Development* **109**, pp. 797–809.

Nornes, S., Clarkson, M., Mikkola, I., Pedersen, M., Bardsley, A., Martinez, J.-P., Krauss, S., and Johansen, T. (1998). Zebrafish contains two *pax6* genes involved in eye development. *Mech Dev* **77**, pp. 185–196.

Northcutt, R. G. (1989). The phylogenetic distribution and innervation of craniate mechanoreceptive lateral lines. In: The Mechanosensory Lateral Line (Coombs, S., Görner, P., and Munz, H., eds.). Springer, New York, pp. 17–78.

Northcutt, R. G. (1996). The origin of craniates: neural crest, neurogenic placodes, and homeobox genes. *Isr J Zool* **42**, pp. 273–313.

Northcutt, R. G. (1997). Evolution of gnathostome lateral line ontogenies. *Brain Behav Evol* **50**, pp. 25–37.

Northcutt, R. G. (2004). Taste buds: development and evolution. *Brain Behav Evol* **64**, pp. 198–206.

Northcutt, R. G., and Gans, C. (1983). The genesis of neural crest and epidermal placodes: a reinterpretation of vertebrate origins. *Q Rev Biol* **58**, pp. 1–28.

Northcutt, R. G., Catania, K. C., and Criley, B. B. (1994). Development of lateral line organs in the axolotl. *J Comp Neurol* **340**, pp. 480–514.

Ogino, H., and Yasuda, K. (1998). Induction of lens differentiation by activation of a bZIP transcription factor, L-Maf. *Science* **280**, pp. 115–118.

Ogino, H., and Yasuda, K. (2000). Sequential activation of transcription factors in lens induction. *Dev Growth Diff* **42**, pp. 437–448.

Ohto, H., Takizawa, T., Saito, T., Kobayashi, M., Ikeda, K., and Kawakami, K. (1998). Tissue and developmental distribution of *Six* family gene products. *Int J Dev Biol* **42**, pp. 141–148.

Ohuchi, H., Hori, Y., Yamasaki, M., Harada, H., Sekine, K., Kato, S., and Itoh, N. (2000). FGF10 acts as a major ligand for FGF receptor 2 IIIB in mouse multi-organ development. *Biochem Biophys Res Commun* **277**, pp. 643–649.

Ohyama, T., and Groves, A. K. (2004). Expression of mouse Foxi class genes in early craniofacial development. *Dev Dyn* **231**, pp. 640–646.

Ohyama, T., Mohamed, O. A., Taketo, M. M., Dufort, D., and Groves, A. K. (2006). Wnt signals mediate a fate decision between otic placode and epidermis. *Development* **133**, pp. 865–875.

Ohyama, T., Groves, A. K., and Martin, K. (2007). The first steps towards hearing: mechanisms of otic placode induction. *Int J Dev Biol* **51**, pp. 463–472.

Oliver, G., Mailhos, A., Wehr, R., Copeland, N. G., Jenkins, N. A., and Gruss, P. (1995a). *Six3*, a murine homologue of the *sine oculis* gene, demarcates the most anterior border of the developing neural plate and is expressed during eye development. *Development* **121**, pp. 4045–4055.

Oliver, G., Sosa-Pineda, B., Geisendorf, S., Spana, E. P., Doe, C. Q., and Gruss, P. (1993). *Prox 1*, a *prospero*-related homeobox gene expressed during mouse development. *Mech Dev* **44**, pp. 3–16.

Oliver, G., Wehr, R., Jenkins, N. A., Copeland, N. G., Cheyette, B. N., Hartenstein, V., Zipursky, S. L., and Gruss, P. (1995b). Homeobox genes and connective tissue patterning. *Development* **121**, pp. 693–705.

Osumi-Yamashita, N., Ninomiya, Y., Doi, H., and Eto, K. (1994). The contribution of both forebrain and midbrain crest cells to the mesenchyme in the frontonasal mass of mouse embryos. *Dev Biol* **164**, pp. 409–419.

Pandur, P. D., and Moody, S. A. (2000). *Xenopus Six1* gene is expressed in neurogenic cranial placodes and maintained in the differentiating lateral lines. *Mech Dev* **96**, pp. 253–257.

Pannese, M., Lupo, G., Kablar, B., Boncinelli, E., Barsacchi, G., and Vignali, R. (1998). The *Xenopus Emx* genes identify presumptive dorsal telencephalon and are induced by head organizer signals. *Mech Dev* **73**, pp. 73–83.

Papalopulu, N. (1995). Regionalization of the forebrain from neural plate to neural tube. *Perspect Dev Neurobiol* **3**, pp. 39–52.

Papalopulu, N., and Kintner, C. (1993). *Xenopus Distal-less* related homeobox genes are expressed in the developing forebrain and are induced by planar signals. *Development* **117**, pp. 961–975.

Park, B.-Y., and Saint-Jeannet, J.-P. (2008). Hindbrain-derived Wnt and Fgf signals cooperate to specify the otic placode in Xenopus. *Dev Biol* **324**, pp. 108–121.

Park, B.-Y., and Saint-Jeannet, J.-P. (2010). Expression analysis of Runx3 and other Runx family members during Xenopus development. *Gene Expr Patterns* **19**, pp. 159–166.

Pattyn, A., Morin, X., Cremer, H., Goridis, C., and Brunet, J.-F. (1997). Expression and interactions of the two closely related homeobox genes *Phox2a* and *Phox2b* during neurogenesis. *Development* **124**, pp. 4065–4075.

Penzel, R., Oschwald, R., Chen, Y., Tacke, L., and Grunz, H. (1997). Characterization and early embryonic expression of a neural specific transcription factor *xSOX3* in *Xenopus laevis*. *Int J Dev Biol* **41**, pp. 667–677.

Pera, E., and Kessel, M. (1999). Expression of DLX3 in chick embryos. *Mech Dev* **89**, pp. 189–193.

Perron, M., Opdecamp, K., Butler, K., Harris, W. A., and Bellefroid, E. J. (1999). X-ngnr-1 and Xath3 promote ectopic expression of sensory neuron markers in the neurula ectoderm and have distinct inducing properties in the retina. *Proc Natl Acad Sci USA* **96**, pp. 14996–15001.

Pfeffer, P. L., Gerster, T., Lun, K., Brand, M., and Busslinger, M. (1998). Characterization of three novel members of the zebrafish *Pax2/5/8* family: dependency of *Pax5* and *Pax8* expression on the *Pax2.1* (*noi*) function. *Development* **125**, pp. 3063–3074.

Phillips, B. T., Bolding, K., and Riley, B. B. (2001). Zebrafish fgf3 and fgf8 encode redundant functions required for otic placode induction. *Dev Biol* **235**, pp. 351–365.

Phillips, B. T., Storch, E. M., Lekven, A. C., and Riley, B. B. (2004). A direct role for Fgf but not Wnt in otic placode induction. *Development* **131**, pp. 923–931.

Pignoni, F., Hu, B., Zavitz, K. H., Xiao, J., Garrity, P. A., and Zipursky, S. L. (1997). The eye-specification proteins So and Eya form a complex and regulate multiple steps in Drosophila eye development. *Cell* **91**, pp. 881–891.

Pissarra, L., Henrique, D., and Duarte, A. (2000). Expression of *hes6*, a new member of the Hairy/Enhancer-of-split family, in mouse development. *Mech Dev* **95**, pp. 275–278.

Pogoda, H.-M., and Hammerschmidt, M. (2009). How to make a teleost adenohypophysis: molecular pathways of pituitary development in zebrafish. *Mol Cell End* **312**, pp. 2–13.

Pohl, B. S., Knöchel, S., Dillinger, K., and Knöchel, W. (2002). Sequence and expression of FoxB2 (XFD-5) and FoxI1c (XFD-10) in Xenopus embryogenesis. *Mech Dev* **117**, pp. 283-287.

Pohl, B. S., Rössner, A., and Knöchel, W. (2005). The Fox gene family in Xenopus laevis: FoxI2, FoxM1 and FoxP1 in early development. *Int J Dev Biol* **49**, pp. 53–58.

Purcell, P., Oliver, G., Mardon, G., Donner, A. L., and Maas, R. L. (2005). Pax6-dependence of Six3, Eya1 and Dach1 expression during lens and nasal placode induction. *Gene Expr Patterns* **6**, pp. 110–118.

Püschel, A. W., Gruss, P., and Westerfield, M. (1992). Sequence and expression pattern of *pax-6* are highly conserved between zebrafish and mice. *Development* **114**, pp. 643–651.

Raible, F., and Brand, M. (2001). Tight transcriptional control of the ETS domain factors Erm and Pea3 by Fgf signaling during early zebrafish development. *Mech Dev* **107**, pp. 105–117.

Ramon-Cueto, A., and Avila, J. (1998). Olfactory ensheathing glia: properties and function. *Brain Res Bull* **46**, pp. 175–187.

Rebay, I., Silver, S. J., and Tootle, T. L. (2005). New vision from Eyes absent: transcription factors as enzymes. *Trends Genet* **21**, pp. 163–171.

Rex, M., Orme, A., Uwanogho, D., Tointon, K., Wigmore, P. M., Sharpe, P. T., and Scotting, P. J. (1997). Dynamic expression of chicken *Sox2* and *Sox3* genes in ectoderm induced to form neural tissue. *Dev Dyn* **209**, pp. 323–332.

Reza, H. M., and Yasuda, K. (2004). Lens differentiation and crystallin regulation: a chick model. *Int J Dev Biol* **48**, pp. 805–817.

Riley, B. B., and Phillips, B. T. (2003). Ringing in the new ear: resolution of cell interactions in otic development. *Dev Biol* **261**, pp. 289–312.

Rinkwitz-Brandt, S., Justus, M., Oldenette, I., Arnold, H. H., and Bober, E. (1995). Distinct temporal expression of mouse *Nkx-5.1* and *Nkx-5.2* homeobox genes during brain and ear development. *Mech Dev* **52**, pp. 371–381.

Rizzoti, K., and Lovell-Badge, R. (2005) Early Development of the Pituitary Gland: induction and Shaping of Rathke's Pouch. *Rev Endocr Metab Disord* **6**, pp. 161–172.

Romand, R., Dolle, P., and Hashino, E. (2006). Retinoic signaling in inner ear development. *J Neurobiol* **66**, pp. 687–704.

Roose, J., Molenaar, M., Peterson, J., Hurenkamp, J., Brantjes, H., Moerer, P., van de Wetering, M., Destree, O., and Clevers, H. (1998). The *Xenopus* Wnt effector XTcf-3 interacts with Groucho-related transcriptional repressors. *Nature* **395**, pp. 608–612.

Ruiz i Altaba, A., and Jessell, T. M. (1991). Retinoic acid modifies the pattern of cell differentiation in the central nervous system of neurula stage Xenopus Embryos. *Development* **112**, pp. 945–958.

Ruvinsky, I., Oates, A. C., Silver, L. M., and Ho, R. K. (2000). The evolution of paired appendages in vertebrates: T-box genes in the zebrafish. *Dev Genes Evol* **210**, pp. 82–91.

Ryan, K., Butler, K., Bellefroid, E., and Gurdon, J. B. (1998). *Xenopus* eomesodermin is expressed in neural differentiation. *Mech Dev* **75**, pp. 155–158.

Sahly, I., Andermann, P., and Petit, C. (1999). The zebrafish *eya1* gene and its expression pattern during embryogenesis. *Dev Genes Evol* **209**, pp. 399–410.

Saint-Germain, N., Lee, Y. H., Zhang, Y., Sargent, T. D., and Saint-Jeannet, J.-P. (2004). Specification of the otic placode depends on Sox9 function in Xenopus. *Development* **131**, pp. 1755–1763.

Sánchez-Calderón, H., Martín-Partido, G., Hidalgo-Sánchez, M. (2002). Differential expression of Otx2, Gbx2, Pax2, and Fgf8 in the developing vestibular and auditory sensory organs. *Brain Res Bull* **57**, pp. 321–323.

Sarrazin, A. F., Nunez, V. A., Sapede, D., Tassin, V., Dambly-Chaudiere, C., and Ghysen, A. (2010). Origin and early development of the posterior lateral line system of zebrafish. *J Neurosci* **30**, pp. in press.

Sasai, Y., and De Robertis, E. M. (1997). Ectodermal patterning in vertebrate embryos. *Dev Biol* **182**, pp. 5–20.

Sbrogna, J. L., Barresi, M. J., and Karlstrom, R. O. (2003). Multiple roles for Hedgehog signaling in zebrafish pituitary development. *Dev Biol* **254**, 19–35.

Schilling, T. F., and Kimmel, C. B. (1994). Segment and cell type lineage restrictions during pharyngeal arch development in the zebrafish embryo. *Development* **120**, pp. 483–494.

Schimmang, T. (2007). Expression and functions of FGF ligands during early otic development. *Int J Dev Biol* **51**, pp. 473–481.

Schlosser, G. (2002a). Development and evolution of lateral line placodes in amphibians. I. Development. *Zoology* **105**, pp. 119–146.

Schlosser, G. (2002b). Development and evolution of lateral line placodes in amphibians. II. Evolutionary diversification. *Zoology* **105**, pp. 177–193.

Schlosser, G. (2003). Hypobranchial placodes in Xenopus laevis give rise to hypobranchial ganglia, a novel type of cranial ganglia. *Cell Tissue Res* **312**, pp. 21–29.

Schlosser, G. (2005). Evolutionary origins of vertebrate placodes: insights from developmental studies and from comparisons with other deuterostomes. *J Exp Zool B Mol Dev Evol* **304B**, pp. 347–399.

Schlosser, G. (2006). Induction and specification of cranial placodes. *Dev Biol* **294**, pp. 303–351.

Schlosser, G. (2007). How old genes make a new head: redeployment of Six and Eya genes during the evolution of the cranial placodes. *Int Comp Biol* **47**, pp. 343–359.

Schlosser, G. (2008). Do vertebrate neural crest and cranial placodes have a common evolutionary origin? *BioEssays* **30**, pp. 659–672.

Schlosser, G., and Northcutt, R. G. (2000). Development of neurogenic placodes in *Xenopus laevis*. *J Comp Neurol* **418**, pp. 121–146.

Schlosser, G., Kintner, C., and Northcutt, R. G. (1999). Loss of ectodermal competence for lateral line placode formation in the direct developing frog *Eleutherodactylus coqui*. *Dev Biol* **213**, pp. 354–369.

Schlosser, G., and Ahrens, K. (2004). Molecular anatomy of placode development in Xenopus laevis. *Dev. Biol.* **271**, pp. 439–466.

Schneider-Maunoury, S., and Pujades, C. (2007). Hindbrain signals in otic regionalization: walk on the wild side. *Int J Dev Biol* **51**, pp. 495–506.

Semina, E. V., Reiter, R. S., and Murray, J. C. (1997). Isolation of a new homeobox gene belonging to the *Pitx/Rieg* family: expression during lens development and mapping to the aphakia region on mouse chromosome 19. *Hum Mol Genet* **6**, pp. 2109–2116.

Semina, E. V., Ferrell, R. E., Mintz-Hittner, H. A., Bitoun, P., Alward, W. L., Reiter, R. S., Funkhauser, C., Daack-Hirsch, S., and Murray, J. C. (1998). A novel homeobox gene *PITX3* is mutated in families with autosomal-dominant cataracts and ASMD. *Nat Genet* **19**, pp. 167–170.

Seo, H. C., Drivenes, Ellingsen, S., and Fjose, A. (1998). Expression of two zebrafish homologues of the murine *Six3* gene demarcates the initial eye primordia. *Mech Dev* **73**, pp. 45–57.

Sharpe, C., and Goldstone, K. (2000). The control of *Xenopus* embryonic primary neurogenesis is mediated by retinoid signaling in the neurectoderm. *Mech Dev* **91**, pp. 69–80.

Sheng, G., and Stern, C. D. (1999). *Gata2* and *Gata3*: novel markers for early embryonic polarity and for non-neural ectoderm in the chick embryo. *Mech Dev* **87**, pp. 213–216.

Sheng, H. Z., and Westphal, H. (1999). Early steps in pituitary organogenesis. *Trends Genet* **15**, pp. 236–240.

Shimeld, S. M., and Holland P. W. H. (2000). Vertebrate innovations. *Proc Natl Acad Sci (USA)* **97**, pp. 4449–4452.

Simeone, A., Gulisano, M., Acampora, D., Stornaiuolo, A., Rambaldi, M., and Boncinelli, E. (1992). Two vertebrate homeobox genes related to the *Drosophila empty spiracles* gene are expressed in the embryonic cerebral cortex. *EMBO J* **11**, pp. 2541–2550.

Sjödal, M., and Gunhaga, L. (2008). Expression patterns of Shh, Ptc2, Raldh3, Pitx2, Isl1, Lim3 and Pax6 in the developing chick hypophyseal placode and Rathke's pouch. *Gene Expr Patterns* **8**, pp. 481–485.

Smotherman, M. S., and Narins, P. M. (2000). Hair cells, hearing and hopping: a field guide to hair cell physiology in the frog. *J Exp Biol* **203**, pp. 2237–2246.

Sjodal, M., Edlund, T., and Gunhaga, L. (2007). Time of exposure to BMP signals plays a key role in the specification of the olfactory and lens placodes ex vivo. *Dev Cell* **13**, pp. 141–149.

Solomon, K. S., Kudoh, T., Dawid, I. B., and Fritz, A. (2003). Zebrafish foxi1 mediates otic placode formation and jaw development. *Development* **130**, pp. 929–940.

Spokony, R. F., Aoki, Y., Saint-Germain, N., Magner-Fink, E. K., and Saint-Jeannet, J.-P. (2002). The transcription factor Sox9 is required for cranial neural crest development in Xenopus. *Development* **129**, pp. 421–432.

Stark, M. R., Sechrist, J., Bronner-Fraser, M., and Marcelle, C. (1997). Neural tube-ectoderm interactions are required for trigeminal placode formation. *Development* **124**, pp. 4287–4295.

Stern, C. D. (2005). Neural induction: old problem, new findings, yet more questions. *Development* **132**, pp. 2007–2021.

Streit, A. (2002). Extensive cell movements accompany formation of the otic placode. *Dev Biol* **249**, pp. 237–254.

Streit, A. (2004). Early development of the cranial sensory nervous system: from a common field to individual placodes. *Dev Biol* **276**, pp. 1–15.

Streit, A. (2007). The pre-placodal region: an ectodermal domain with multipotential progenitors that contribute to sense organs and cranial sensory ganglia. *Int J Dev Biol* **51**, pp. 447–461.

Streit, A., and Stern, C. (1999). Establishment and maintenance of the border of the neural plate in the chick: involvement of FGF and BMP activity. *Mech Dev* **82**, pp. 51–66.

Streit, A., Berliner, A. J., Papanayotou, C., Sirulnik, A., and Stern, C. D. (2000). Initiation of neural induction by FGF signalling before gastrulation. *Nature* **406**, pp. 74–78.

Sun, S. K., Dee, C. T., Tripathi, V. B., Rengifo, A., Hirst, C. S., and Scotting, P. J. (2007). Epibranchial and otic placodes are induced by a common Fgf signal, but their subsequent development is independent. *Dev Biol* **303**, pp. 675–686.

Swindell, E. C., Zilinski, C. A., Hashimoto, R., Shah, R., Lane, M. E., and Jamrich, M. (2008). Regulation and function of foxe3 during early zebrafish development. *Genesis* **46**, pp. 177–183.

Taira, M., Hayes, W. P., Otani, H., and Dawid, I. B. (1993). Expression of LIM class homeobox gene *Xlim-3* in *Xenopus* development is limited to neural and neuroendocrine tissues. *Dev Biol* **159**, pp. 245–256.

Takabatake, Y., Takabatake, T., and Takeshima, K. (2000). Conserved and divergent expression of T-box genes *Tbx2-Tbx5* in *Xenopus. Mech Dev* **91**, pp. 433–437.

Taniguchi, K., Saito, S., Oikawa, T., and Taniguchi, K. (2008). Phylogenic aspects of the amphibian dual olfactory system. *J Vet Med Sci* **1**, pp. 1–9.

Tiveron, M. C., Hirsch, M. R., and Brunet, J. F. (1996). The expression pattern of the transcription factor Phox2 delineates synaptic pathways of the autonomic nervous system. *J Neurosci* **16**, pp. 7649–7660.

Tomarev, S. I., Sundin, O., Banerjee-Basu, S., Duncan, M. K., Yang, J. M., and Piatigorsky, J. (1996). Chicken homeobox gene *Prox 1* related to *Drosophila prospero* is expressed in the developing lens and retina. *Dev Dyn* **206**, pp. 354–367.

Tongiorgi, E., Bernhardt, R. R., and Schachner, M. (1995). Zebrafish neurons express two L1-related molecules during early axonogenesis. *J Neurosci Res* **42**, pp. 547–561.

Torres, M., and Giráldez, F. (1998). The development of the vertebrate inner ear. *Mech Dev* **71**, pp. 5–21.

Torres, M., Gómez-Pardo, E., and Gruss, P. (1996). Pax2 contributes to inner ear patterning and optic nerve trajectory. *Development* **122**, pp. 3381–3391.

Tracey, W. D., Pepling, M. E., Marko, E. H., Thomsen, G. H., and Gerben, J.-P. (1998). A Xenopus homologue of aml-1 reveals unexpected patterning mechanisms leading to the formation of embryonic blood. *Development* **125**, pp. 1371–1380.

Treier, M., Gleiberman, A. S., O'Connell, S. M., Szeto, D. P., McMahon, J. A., McMahon, A. P., and Rosenfeld, M. G. (1998). Multistep signaling requirements for pituitary organogenesis *in vivo. Genes Dev* **12**, pp. 1691–1704.

Tremblay, P., Kessel, M., and Gruss, P. (1995). A transgenic neuroanatomical marker identifies cranial neural crest deficiencies associated with the Pax3 mutant Splotch. *Dev Biol* **171**, pp. 317–329.

Tribulo, C., Ayba, M. J., Nguyen, V. H., Mullins, M. C., and Mayor, R. (2003). Regulation of Msx genes by Bmp gradient is essential fro neural crest specification. *Development* **130**, pp. 6441–6452.

Tripathi, V. B., Ishii, Y., Abu-Elmagd, M. M., and Scotting, P. J. (2009). The surface ectoderm of the chick embryo exhibits dynamic variation in its response to neurogenic signals. *Int J Dev Biol* **53**, pp. 1023–1033.

Vacalla, C. M., and Theil, T. (2002). Cst, a novel mouse gene related to Drosophila Castor, exhibits dynamic expression patterns during neurogenesis and heart development. *Mech Dev* **118**, pp. 265–268.

Valarché , I., Tissier-Seta, J.-P., Hirsch, M. R., Martinez, S., Goridis, C., and Brunet, J. F. (1993). The mouse homeodomain protein Phox2 regulates *Ncam* promoter activity in concert with Cux/ CDP and is a putative determinant of neurotransmitter phenotype. *Development* **119**, pp. 881–896.

Vasiliauskas, D., and Stern, C. D. (2000). Expression of mouse *HES-6*, a new member of the Hairy/ Enhancer of split family of bHLH transcription factors. *Mech Dev* **98**, pp. 133–137.

Wada, H., Saiga, H., Satoh, N., and Holland, P. W. H. (1998). Tripartite organization of the ancestral chordate brain and the antiquity of placodes: insights from ascidian Pax-2/5/8, Hox and Otx genes. *Development* **125**, pp. 1113–1122.

Wall, N. A., Jones, C. M., Hogan, B. L. M., and Wright, C. V. E. (1992). Expression and modification of Hox 2.1 protein in mouse embryos. *Mech Dev* **37**, pp. 111–120.

Walther, C., and Gruss, P. (1991). *Pax-6*, a murine paired box gene, is expressed in the developing CNS. *Development* **113**, pp. 1435–1449.

Wang, W., Van De Water, T., and Lufkin, T. (1998). Inner ear and maternal reproductive defects in mice lacking the *Hmx3* homeobox gene. *Development* **125**, pp. 621–634.

Wawersik, S., Purcell, P., Rauchman, M., Dudley, A. T., Robertson, E. J., and Maas, R. (1999). BMP7 acts in murine lens placode development. *Dev Biol* **207**, pp. 176–188.

Webb, J. F. (1989) Gross morphology and evolution of the mechanoreceptive lateral-line system in teleost fishes. *Brain Behav Evol* **33**, pp. 34–53.

Webb, J. F., and Noden, D. M. (1993). Ectodermal placodes: Contributions to the development of the vertebrate head. *Am. Zool.* **33**, 434–447.

West-Mays, J. A., Zhang, J., Nottoli, T., Hagopian-Donaldson, S., Libby, D., Strissel, K. J., and Williams, T. (1999). AP-2a transcription factor is required for early morphogenesis of the lens vesicle. *Dev Biol* **206**, pp. 46–62.

Whitfield, T. T., Granato, M., van Eeden, F. J., Schach, U., Brand, M., Furutani- Seiki, M., Haffter, P., Hammerschmidt, M., Heisenberg, C. P., Jiang, Y. J., Kane, D. A., Kelsh, R. N., Mullins, M. C., Odenthal, J., and Nüsslein-Volhard, C. (1996). Mutations affecting development of the zebrafish inner ear and lateral line. *Development* **123**, pp. 241–254.

Whitfield, T. T., Riley, B. B., Chiang, M. Y., and Phillips, B. (2002). Development of the zebrafish inner ear. *Dev Dyn* **223**, pp. 427–458.

Whitlock, K. E. (2005). Origin and development of GnRH neurons. *Trend Endocrinol Metab* **16**, pp. 145–151.

Whitlock, K. E., Wolf, C. D., and Boyce, M. L. (2003). Gonadotropin-releasing hormone (GnRH) cells arise from cranial neural crest and adenohypophyseal regions of the neural plate in the zebrafish, Danio rerio. *Dev. Biol.* **257**, pp. 140–152.

Wigle, J. T., Chowdhury, K., Gruss, P., and Oliver, G. (1999). *Prox1* function is crucial for mouse lens-fibre elongation. *Nat Genet* **21**, pp. 318–322.

Wilson, P. A., Lagna, G., Suzuki, A., and Hemmati-Brivanlou, A. (1997). Concentration-dependent patterning of the Xenopus ectoderm by BMP4 and its signal transducer Smad1. *Development* **124**, pp. 3177–3184.

Winklbauer, R. (1989). Development of the lateral line system in *Xenopus Prog Neurobiol* **32**, pp. 181–206.

Woo, K., and Fraser, S. E. (1998). Specification of the hindbrain fate in zebrafish. *Dev Biol* **197**, pp. 283–296.

Wood, H. B., and Episkopou, V. (1999). Comparative expression of the mouse *Sox1*, *Sox2* and *Sox3* genes from pre-gastrulation to early somite stages. *Mech Dev* **86**, pp. 197–201.

Wray, S. (2002). Molecular mechanisms for migration of placodally derived GnRH neurons. *Chem. Sense* **27**, pp. 569–572.

Wright, T. J., and Mansour, S. L. (2003). Fgf3 and Fgf10 are required for mouse otic placode induction. *Development* **130**, pp. 3379–3390.

Wurm, A., Sock, E., Fuchshofer, R., Wegner, M., and Tamm, E. R. (2008). Anterior segment dysgenesis in the eyes of mice deficient for the high-mobility-group transcription factor Sox11. *Exp Eye Res* **86**, pp. 895–907.

Xu, P. X., Woo, I., Her, H., Beier, D. R., and Maas, R. L. (1997). Mouse *Eya* homologues of the *Drosophila eyes absent* gene require *Pax6* for expression in lens and nasal placode. *Development* **124**, pp. 219–231.

Xu, P. X., Adams, J., Peters, H., Brown, M. C., Heaney, S., and Maas, R. (1999). Eya1-deficient mice lack ears and kidneys and show abnormal apoptosis of organ primordia. *Nat Genet* **23**, pp. 113–117.

Xu, H., Dude, C. M., and Baker, C. V. (2008). Fine-grained fate maps for the ophthalmic and maxillomandibular trigeminal placodes in the chick embryo. *Dev Biol* **317**, pp. 174–186.

Yang, L., Zhang, H., Hu, G., Wang, H., Abate-Shen, C., and Shen, M. M. (1998). An early phase of embryonic *Dlx5* expression defines the rostral boundary of the neural plate. *J Neurosci* **18**, pp. 8322–8330.

Yoshimoto, A., Saigou, Y., Higashi, Y., and Kondoh, H. (2005). Regulation of ocular lens development by Smad-interacting protein 1 involving Foxe3 activation. *Development* **132**, pp. 4437–4448.

Zelarayan, L. C., Vendrell, V., Alvarez, Y., Dominguez-Frutos, E., Theil, T., Alonso, M. T., Maconochie, M., and Schimmang, T. (2007). Differential requirements for FGF3, FGF8 and FGF10 during inner ear development. *Dev Biol* **308**, 379–391.

Zhang, X., Friedman, A., Heaney, S., Purcell, P., and Maas, R. L. (2002). Meis homeoproteins directly regulate Pax6 during vertebrate lens morphogenesis. *Genes Dev* **16**, pp. 2097–2107.

Zheng, W., Huang, L., Wei, Z. B., Silvius, D., Tang, B., and Xu, P. X. (2003). The role of Six1 in mammalian auditory system development. *Development* **130**, pp. 3989–4000.

Zilinski, C. A., Shah, R., Lane, M. E., and Jamrich, M. (2005). Modulation of zebrafish pitx3 expression in the primordia of the pituitary, lens, olfactory epithelium and cranial ganglia by hedgehog and nodal signaling. *Genesis* **41**, pp. 33–40.

Zlotnik, A., and Yoshie, O. (2000). Chemokines: a new classification system and their role in immunity. *Immunity* **12**, pp. 121–127.

Zou, D., Silvius, D., Fritzsch, B., and Xu, P. X. (2004). Eya1 and Six1 are essential for early steps of sensory neurogenesis in mammalian cranial placodes. *Development* **131**, pp. 5561–5572.

Zygar, C. A., Cook, T. L. J., and Grainger, R. M. (1998). Gene activation during early stages of lens induction in *Xenopus laevis Development* **125**, pp. 3509–3519.

Author Biographies

Jean-Pierre Saint-Jeannet was born in Mirande and grew up in the small town of L'isle de Noe (Gers, France). Jean-Pierre received is education in France. He obtained a Ph.D. in Developmental Neurobiology from Paul Sabatier University in Toulouse where he worked with Dr. Anne-Marie Duprat on the mechanisms underlying neural induction in the salamander *Pleurodeles waltlii*. He subsequently trained as a postdoctoral fellow in the laboratory of Dr. Jean Paul Thiery at the Ecole Normale Superieure in Paris, studying cadherin-mediated cell-cell adhesion during development. Jean-Pierre then performed a second postdoc with Dr. Igor B. Dawid at the National Institutes of Health in Bethesda studying the role of canonical Wnt signaling pathway in neural crest formation. Jean-Pierre is currently a Chargé de Recherche at the CNRS and an Associate Professor of Developmental Biology in the Department of Animal Biology at the University of Pennsylvania, School of Veterinary Medicine in Phildalephia. Research in his laboratory focuses on the molecular mechanisms controlling neural crest and placodes formation using *Xenopus laevis* as a model system.

Byung-Yong Park was born in Jeonju (South Korea) where he grew up. He did his graduate education at Chonbuk National University where he received his Veterinary and PhD degrees. As part of his graduate work Byung-Yong studied the mechanisms of mouse craniofacial development in the laboratory of Dr. In-Shik Kim. After graduation Byung-Yong joined the laboratory of Dr. Jean-Pierre Saint-Jeannet at the University of Pennsylvania, to study the factors regulating otic placode development in Xenopus. Byung-Yong recently accepted a position in the Department of Anatomy, College of Veterinary Medicine at Chonbuk National University in Korea where he will continue his work on placode induction and differentiation. Byung-Yong is the proud father of two children Kunook (Kevin) and Sayun (Helen).